Production Technology of Spices, Aromatic Medicinal and Plantation Crops

Production Technology of Spices, Aromatic Medicinal and Plantation Crops

Swati Barche
Associate Professor (Horticulture)
College of Agriculture, Indore
R.V.S.K.V.V., Gwalior, M.P.

Kamal S. Kirad
Senior Scientist and Head
KVK, R.V.S.K.V.V., Dhar, M.P.

Reena Nair
P.K. Jain
S.K. Sengupta
College of Agriculture
JNKVV, Adhartal, Jabalpur - 482004, M.P.

NEW INDIA PUBLISHING AGENCY
New Delhi – 110 034

NEW INDIA PUBLISHING AGENCY
101, Vikas Surya Plaza, CU Block, LSC Market
Pitam Pura, New Delhi 110 034, India
Phone: + 91 (11)27 34 17 17 Fax: + 91(11) 27 34 16 16
Email: info@nipabooks.com
Web: www.nipabooks.com

Feedback at feedbacks@nipabooks.com

ISBN: 978-93-85516-06-1

Composed and Designed by NIPA

Preface

India's is bestowed with diverse agro-climatic and edaphic conditions and grow vast range of different agricultural crops. Among these, Horticulture share is 33.35 percent in our agricultural GDP. It includes not only fruit, vegetable, flowers but also spices, Plantation, medicinal, aromatic, Bamboo plants spices and plantation crops play a key role in export and fetches higher income for country. Medicinal and aromatic plants are highly valuable as they are the key source for the indigenous pharmaceutical, perfumery, flavour and cosmetic industry. Country is also earning foreign exchange from these group of plants. Keeping in view the importance of spices, plantation, medicinal and aromatic crops the book has been compiled and the information gathered is percolated form various sources.

No doubt a few books on the subject are available but have scattered form. The valuable research, extension literature is available in the various annual reports, symposia-seminar proceedings and students of UG and PG are unable to avail these literature.

In such a situation the need for a simple book on these low volume high crops is highly felt by the students of Agriculture, Horticulture or Forestry as majority of Agricultural Universities of India and have these crops in their curriculum. This book is thus broughtout mainly to meet the above requirement & is also the outcome of our practical experience.

Hope, this book will be a boon for students, researchers, extension workers and commercial growers.

Authors

Contents

1

Introduction

Spices are non-leafy parts (*e.g.* bud, fruit, seed, bark, rhizome and bulb) of plants used as a flavoring or seasoning, although many can also be used as a herbal medicine. They are those plants, the products of which are made use as food adjuncts to add aroma and flavour or they are derived vegetable substances which are used for seasoning the food imparting distinctive flavour and fragrance that gives zest or piquancy for enjoyment. The term 'spice' originated from the Latin word '*species*', meaning of specific kind. A closely related term, 'herb', is used to distinguish plant parts finding the same uses but derived from leafy or soft flowering parts. For example Black pepper, Ginger, Turmeric, Chilli, Cinnamon, Nutmeg etc. Spices are generally composed of fiber, carbohydrate, fat, sugar, protein, gum, ash, volatile (essential oils), and other nonvolatile components. All of these components impart each spice's particular flavour, colour, nutritional, health, or preservative effects. The flavour components (volatile and nonvolatile) are protected within a matrix of carbohydrate, protein, fiber, and other cell components. When the spice is ground, cut, or crushed, this cell matrix breaks down and releases the volatile components (Raghavan, 2007).

Condiments are the plant products which add taste to the food stuff e.g., Coriander. The term plantation crops refers to those tropical crops which are cultivated on an extensive scale in a large contiguous area, owned and managed by an individual or a company or grown on a large scale by a uniform system of cultivation under central management e.g., Tea, Coffee, Rubber, Cocoa, Coconut, Arecanut, Oil palm, Palmyrah palm, Cashewnut etc. In a modern science, the approach in growing of plantation crops is closer to a factory than to a farm which stimulates and encourages even the small holders to send their produce to the factory for processing.

These crops are of high commercial value having great economic importance. Thus, they play a vital role in the agricultural economy of the country. Among the horticultural crop production, spices and plantation contributes 1.73 percent and 5.17 percent respectively. Major spices producing states are Andhra Pradesh (19%), Gujarat (15%), Rajasthan (14.7%), Karnataka (8%), Madhya Pradesh (7.7%) and Tamil Nadu (7%). The spice production in India is currently estimated at 5.95 million tonnes from an area of about 3.21 million ha. The area, production and productivity of spices, condiments and plantation crops in India is given in Appendix-I. Plantation crops are largely grown in the tropics between 20° North and 20° South Latitude of equator. Most of the plantation crops are perennial trees while some are climbers and shrubs. They are high-value commercial crops having great economic importance and play vital role in export trade. Major plantation crops producing states are Tamil Nadu (28%), Karnataka (26%), Kerala (25.5%), Andhra Pradesh (8%) which all together contributes about 87.5% of the plantation crop production in the country.

India has been considered as treasure house of a large number of valuable medicinal and aromatic plant species. Medicinal and aromatic plants form a numerically large group of economically important plants which provide basic raw materials for medicines, perfumes, flavours and cosmetics. In Indian tradition, all plants on earth are considered as medicinal, however, a simplest definition of medicinal plant is that the plants which are used in official and various traditional system of medicine throughout the world or These are plants that provides people with medicines to prevent disease, maintain health or cure ailment. The basis for this use is that these plants contain so called active ingredients (active principles or biologically active principles) that affect physiological (metabolic) processes of living organisms.

Aromatic plants are defined as those plants possess essential oil in them. This essential oil is the odoriferous steam volatile constituents of the aromatic plants. They are mainly a complex mixture of cyclic and or cyclic monoterpenoids. These terpenoids are basically secondary metabolites and they have no apparent function in the plant metabolism. Aromatic plants possess odorous volatile substances which occur as essential oil, gum exudate, balsam and oleoresin in one or more parts, namely, root, wood, bark, stem, foliage, flower and fruit. The characteristic aroma is due to a variety of complex chemical compounds.

The pharmacological activity of medicinal and aromatic plants (MAPs) is due to their biologically active ingredients classified into four groups: alkaloids, glycosides, essential oils and other miscellaneous active substances. Secondary metabolites being end products of metabolic processes can be observed through colour, taste or odour and is determined by chemical analysis.

Importance and scope of spices

Spices constitute an important group of agricultural commodities, which since antiquity have been considered indispensable in the culinary art for flavoring of foods. India is known as the land of spices and is the largest producer, consumer and exporter of spices and spice products in the world. Over 100 plant species are known to yield spices and spice products among which around 50 are grown in India. The varied agro climatic situations suits for almost all spices. The production of spices in the country has registered a substantial increase over the last ten years with average annual growth of 5.8%. Chilli is the major spice crop occupying about 25% of area under cultivation and contributing 22% of total spice production in the country. Garlic accounts for 8.0% of area with 21.0% share in production, while turmeric accounts for 6.8% of area with 19.6% share in production. India recorded an increase of 22 per cent in the export of spices and spice-based products during 2012-13 to touch 699,170 tonnes, as against 575,270 tonnes in the previous financial year.

Our country is suitable for production of tropical black pepper to temperate saffron spices and herbs have tremendous importance in every day life, as ingredients, in food, alcoholic beverages, medicine colouring and as gardening plants. They constitute a predominant sector in earning valuable foreign exchange and is considered as a cash crop. Some of the spices are also used in pharmaceuticals, perfumes, cosmetics and several other industries. They contain essential oils, which provide the flavour and taste. Some important flavouring compounds from major spices is given in Appendix-II.

The first spice oil and oleoresin industry was started in 1930 in India at Calicut by a private entrepreneur. Extracts of ginger were manufactured during the Second World War. The major oils are from black pepper, cardamom, chilli seed, paprika, clove, nutmeg, mace, cinnamon, cassia, kokkam, galangal, juniper and peppermint (Guenther, 1950). Pepper oil, ginger oil, celery seed oil, kokkam oil and peppermint are the major oils exported from India.

Turmeric is a good source of natural colours, having curcumin which has became an important part in food, preservatives and body care industry. The increasing demand for natural colours can also be exploited from Indian paprika, which has high colour value. Different spices have different property, like the anti oxidant property of curcumin, anti-obesity factor from garcinia, piperin from pepper, zingiberin from ginger, anti-cholestrol and anti diabetic properties of garlic, onion and fenugreek can be exploited. Spices act as powerful antibiotics. Onion, garlic and allspice kill the bacteria and they were effectively tested against *Salmonella* and *Staphylococcus* a food poisoning bacterium. Other spices such as hot peppers destroys at least 75% of their bacterial targets. Cloves are used as preservative food and inhibits the growth of bacteria.

Constraints in spices production

The problem in successful production varies with location and crop. In general following are some of the constraints in production.

i. Inadequate supply of quality seed/ planting material of improved varieties.

ii. Lack of appropriate inputs other than seed and their timely availability.

iii. Climate and weather variations during production.

iv. Pests and diseases.

v. Lesser fertilizer inputs, low fertilizer use efficiency and micronutrient deficiency.

vi. Market fluctuations.

vii. Poor post harvest handling, processing and inadequate storage facilities.

Strategies to improve productivity of spices

Though India has been acclaimed as the land of spices, Indian spices industry has been facing a stiff competition from other spices producing countries. There has been an alarming growth in the consumption pattern of spices in India itself. Indian spices are known for their intrinsic quality, flavour and aroma. The strong domestic market is a great booster for augmenting production of spices. As the international market becomes increasingly competitive, it is imperative to enhance the production and bring down cost of cultivation to make the Indian spices globally competitive. India can sustain production and recapture the international market only by attaining reduction in unit cost of production by increasing productivity.

Low productivity and high production cost make Indian spices costlier in the world market. Several reasons like existence of uneconomic and senile genetic stock, incidence of serious diseases like phytophthora foot rot in pepper, viral disease in cardamom, soft rot and bacterial wilt in ginger are attributed to low productivity of spices in our country. There is an urgent need to develop high yielding disease free varieties and encourage multistory cropping and inter cropping to maximize the yields

The research and development stages of spices are very strong in India. A number of research organizations, *viz* Indian Institute of Spices Research (IISR), Calicut, Kerela, National Research Centre for Seed Spices, Ajmer, Rajasthan, Spices Board, Indian Cardamom Research Institute, (ICRI), Myladumpara, Kerela, Central Food Technological Research Laboratory, Mysore, Karnataka, United Planters Association of South India (UPASI) and a number of State

Agricultural Universities (SAUs) in India are playing an important role in evolving varieties, agrotechniques, plant protection measures, value additions and transfer of technologies on spices.

The Spices Board under Ministry of Commerce , Government of India with its head quarters at Cochin, Directorate of Spices and Arecanut, Calicut, Kerela and a number of Horticulture and Agriculture departments and Non Governments Organizations (NGOs) have been playing pivotal role in development, promotion of spices and export.

Importance of plantation crops

There are two groups of plantation crops namely coconut, arecanut, oil palm, cocoa, cashewnut dealt by Ministry of Agriculture and tea, coffee, rubber dealt by the Ministry of Commerce. These crops produce higher biomass than field crops per unit area resulting in efficient utilization of natural resources. It also has potential to improve wastelands. They help in alleviation of poverty level in varied agro- ecosystems like rainfed, dry land, hilly, arid and coastal. Provide higher employment opportunities and are important for nutritional security. They have high potential for value-addition and foreign exchange.

Coconut, the "tree of Heaven" or *Kalpavriksha* has many uses as food, edible oil, industrial oil for making various products, as a healthy drink, *etc*. the timber, leaf, petiole, shell, husk of coconut are useful for various purposes. Arecanut is primarily used as mastigatory and as a stimulant with betel leaf and also as pan masala, scented supari, *etc*. Oil palm is the highest oil yielding perennial crop and yields two types of oil *viz*., palm oil and kernel oil. Palm oil is rich in vitamin A and E. It is also being utilized in diets to overcome vitamin A deficiency in children. Palm oil is a source of nutrition and health due to its high calorific value. It is readily digestible and easily absorable. It contains equal proportion of saturated and unsaturated fatty acids.

India is the largest producer, processor, consumer and exporter of cashew in the world. Cashewnut was considered as a poor man's crop till fifties, but recently attained higher commercial importance. Cashew apple fenni is a popular alcoholic drink in Goa. Cashewnut shell liquid is industrial oil. Cocoa beans are largely used in the confectionery industrial sector for which demand is on increase. The high confectionery value of cocoa beans derived after fermentation paves way for the manufacture of cocoa butter and cocoa powder which ultimately becomes the raw material for manufacture of edible chocolates, drinking chocolates, *etc*. The health benefit of tea was realized in 273BC by the Chinese Emperor Shew Mung. He was convinced that tea improved health and recommended it as remedy for kidney trouble, fever, chest infection and tumors in the head. It is a dietary polyphenol and good source of fluoride to improve oral health.

Most of the rubber plantations are located in far-flung villages and they play a major role in rural development of the country. Women are employed in plenty in rubber plantations and this contributes to their empowerment. In employment generation, rubber-manufacturing industry also play an equally important role. Though rubber is traditionally grown in the southern states of Kerala and parts of Tamil Nadu and Karnataka, the crop has now gained acceptance among the people of non-traditional regions as well, especially the North-east. Rubber plantations are very helpful in protecting the environment.

Constraints in plantation production

i. Old, senile and unproductive plantation as well as gaps in the existing plantation.

ii. Unavailability of quality planting material.

iii. Insect/ pest and diseases problem.

iv. High initial investments and long gestation period.

v. Inadequate on-farm production handling/ processing suitable for export purpose.

vi. Poor marketing structure, lack of market research studies.

vii. Lack of premium price to the farmer for quality produce.

viii. Poor post harvest technology.

ix. Lack in modern processing facilities.

x. Inadequate transfer of technology mechanism.

xi. Inadequate human resource developments.

Development strategies

i. Competitiveness by increasing the production and productivity by adoption of newer technologies.

ii. Making available of genetically pure planting material.

iii. Popularization of vegetative propagated planting material wherever technologies are available.

iv. Popularization of intercropping and mix farming, particularly for coconut, arecanut, coffee, tea, and rubber.

v. Providing entrepreneurship skill and micro credit to start the processing.

vi. Exploitation of ecofriendly nature of the plantation crops for a better environment.

vii. Promoting farmers' participating approach in organizing training and demonstration.

Importance of medicinal and aromatic crops

Medicinal and Aromatic plants are one of the major natural resources of our country occurring in diverse ecosystems. India and China are the two largest countries in the world which have the richest traditional system of medicine such as *Ayurveda, Siddha,* Chinese medicine etc. which uses large number of aromatic and medicinal plants. North East India is one of the richest repositories of these plants in the world which will pave way for reaping the Green Gold. India is a veritable emporium of essential oil bearing plants embracing perfumery as one of the Indian arts and our country was a famous centre for perfumers and allied products in the past for a long time, a flourishing perfumery industry existed in India, and rose and sandalwood based attars were even being exported. The world trend in essential oil is vast.

Although, India's share in the growing world herbal market is negligible mainly because of inadequate investment in this sector in terms of research and validation of old heritage knowledge in the light of modern science that is acceptable to the western world. However, potential of India is immense, if cultivars efforts are put forth with a clear vision to capture a lion's share of the world herbal trade in coming years. We need to give special emphasis on natural products derived from aromatic plants from rich biodiversity of existing areas.

It provides natural source of high value industrial raw material for pharmaceutical, agricultural chemicals, food and cosmetic industries and opens up new possibilities for higher level of gains for farmers with a significant scope for progress in rural economy. Essential oils are highly concentrated, low volume, high value products. Essential oils or value added extracts or products required for perfumery, flavour, fragrance and cosmetic industries in addition to aromatherapy value are other important areas where India can contribute to business and demand globally. It ensures the production of assured and uniform quality material and can be act as economic driven participatory conservation.

Nature has been a source of medicinal agents' for thousands of years and an impressive number of modern drugs have been isolated from natural sources, many based on their use in traditional medicine. The widespread use of herbal remedies and health care preparations, such as those described in ancient texts like the Vedas, has been traced to the occurrence of natural products with medicinal properties. In fact, plants produce a diverse range of bioactive molecules, making them a rich source of different types of medicines. Higher plants, as sources of medicinal compounds, have continued to play a dominant role in the maintenance of human health since ancient times. Over 50% of all modern clinical drugs are of natural product origin and natural products play an important role in drug development programs in the pharmaceutical industries.

Constraints in medicinal and aromatic plant production

1. Information on propagation of Medicinal and Aromatic plants is available only for less than 10% and agro techniques for only 1% of the total plants globally.
2. In order to fulfill the increasing demand of herbal drugs Aromatic plants farming in large areas is imperative.

Development strategies

Since Indian subcontinent is well known for its age old healthcare traditions there is an urgent need to establish these traditional know how in both the national and international perspective realizing the ongoing development trends in traditional knowledge. Aromatic plants are essentially important as alternate income generating sources to the underprivileged communities dwelling in forest. Therefore strengthening of this sector may benefit and improve the living standards of farmers as well as forest dwellers. Herbal gardens are being promoted to lay emphasis on sensitization of MAP (Medicinal and aromatic plants) cultivation. Any meaningful programme would call for commercializing those plants that are required in large quantities in domestic and international markets. There is a need to establish a regulatory and controlling authority on the lines of drug and cosmetics Act and Seed & Nursery Act to prevent unscrupulous activity and boost the image of Indian MAP sector. Establishment of referral labs for crude drug analysis to ensure authenticity of herbal raw material. Priority needs to be given in areas of new drug development from plants, development of pharmacopeia standards for quality, safety, efficacy and scientific validation of therapeutic claims. A strong and viable national strategy is warranted for complete inventorisation, economic mapping, conservation, cultivation, post harvest technology and storage facilities of MAP crops.

References

Guenther, E. 1950. The Essential oils, Volume IV Von Nostrond Publishing Co., New York, pp 396-437.

Raghavan, S. 2007. Handbook of spices, seasonings, and flavorings. 2nd Ed. CRC Press, Taylor & Francis Group, Boca Raton.

State of IndianAgriculture 2012-2013. GOI, Ministry of Agriculture, Department of Agricuture and Coorporation. pp 72-98.

Report of the Working Group on Horticulture, Plantation Crops and Organic Farming for the XI Five Year Plan (2007-12) GOI, Planning Commission, January 2007. pp 1-40.

2

Production Technology of Spices

2.1 GINGER

Botanical Name	*Zingiber officinale* Roscoe
Family	Zingiberaceae
Origin	South East Asia, China
Chromosome number	2n=22, 96
Commercial value	Dried rhizome
Chemical constituents	The major pungent principle in ginger oleoresin is gingerol
Research Institute	Indian Institute of Spice Research, Calicut, Kerala

Importance

Ginger is a perennial herbaceous monocotyledon, usually grown as annual, and is known to human as spice as well as medicinal crop. The word *Zingiber* is thought to come from the Sanskrit word *singabera* meaning 'shaped like a horn'. It probably got its name because the rhizomes look like deer's antlers. It first appeared in the writings of Confucius in the 5th century B.C.

The economic part is the underground rhizomes, which is pungent and aromatic. It is valued for its medicinal properties and is useful against gastritis, dyspepsia, cold and cough. It is widely used in food, beverages, confectionery, etc. Baby ginger or young ginger results from early harvesting and is a gourmet quality product with a tender, low-fiber texture and is mostly used for pickling. Rhizome contains proteins, fibre, starch, essential oil, Zingiberol, Glutamic acid, Aspartic acid, Glucose, Fructose, and Sucrose. Most ginger varieties have about 3-6% crude fibre, 1.5-3.0% volatile oil and 4-10% oleoresin. The major pungent principle in ginger oleoresin is gingerol. Among the gingerols 6-gingerol, 8-gingerol and 10-gingerol are very important in contributing the pungency. Gingerol (Hydroxy methoxy phenyl decanone) on storage gets converted to another compound called shogaol. The age of ginger oleoresin can be assessed based on the shogaol content. Chemically, gingerol is a relative of capsaicin, the compound that gives chilli its spiciness. Cooking ginger transforms gingerol into zingerone, which is less pungent and has a spicy-sweet aroma.

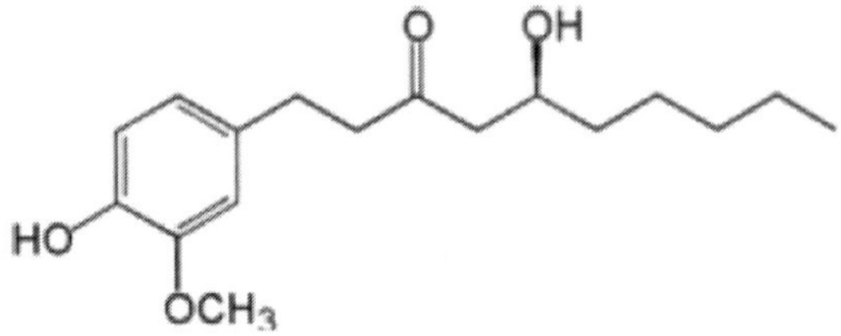

Gingerol ($C_{17}H_{26}O_4$)

Distribution

Ginger has been grown in India and China since ancient times, and by the first century traders has been brought it to the Mediterranean region. Obtained by the Greeks and Romans from Arab traders, it was one of the first oriental spices to arrive in Europe. Today ginger has been grown in most of the warmer part of the world. Hawaii and Fiji are the major ginger producing countries in the Pacific. It is cultivated in most of the states in India. However, states namely Kerala, Meghalaya, Arunachal Pradesh, Mizoram, Sikkim, Nagaland and Orissa together contribute 70 per cent to the country's total production.

Description of plant

Ginger plants can grow to about 1 m tall. The upright shoots sprout from the rhizome at the base of the plant. The shoots (pseudostem), grow up to 1.2 m tall, arise annually from buds on the rhizome. These pseudostems are formed from a series of leaf bases (sheaths) wrapped tightly around one another with the long (up to 7 cm), narrow (up to 1.9 cm wide), mid-green leaf blades arranged alternately. The flowering heads, borne on separate shorter stems, are cone-shaped spikes and composed of a series of greenish to yellowish leaf-like bracts. Protruding just beyond the outer edge of the bracts, the flowers are pale yellow in colour with a purplish tip that has yellowish dots and striations. Flowering stems are rarely, produced in cultivated plants. Rhizomes are knobby and fleshy, covered in ring-like scars. This is the important part for food and medicine. Although the rhizomes grow underground, they are swollen stems, not roots. This is why fresh ginger is often referred to as 'stem ginger'.

Soil

It can be cultivated in almost all types of soils but it thrives best in deep (25-40 cm), rock free, well drained sandy or clay loam, red loam or lateritic loam soils. By following bed system of planting and surface mulching, shallower soils can be utilized for cultivation. A friable loam rich in humus is ideal. It is an exhausting crop so it may not be desirable to grow ginger in same site year after year. The pH of 6-7 is ideal for its cultivation.

Climate

Ginger is a tropical crop, it requires warm and humid climate. It is cultivated from sea level to an altitude of 1500 meters above sea level. For successful cultivation of ginger suitable temperature ranges from 25-28° C. Growth efficiency declines with temperature above 30° C and below 24° C. It can be grown in full sun as well as partial shade when grown as intercrop. Vegetative growth is promoted under long day lengths and rhizome enlargement is promoted under shorter day lengths. A rainfall of 1500-3000mm well distributed for 8-10 months is ideal. Moderate rainfall at the sowing time till the rhizomes sprout followed by fairly heavy and well distributed showers during the growing period and dry weather for about a month before harvesting is required for large scale cultivation.

Cultivation

Sowing time

In coastal region of India, April till first fortnight of May with the receipt of pre monsoon showers is the best time for planting. If irrigation facilities are available then it can be planted well in advance during the middle of February or early March. Early planting can reduce the infestation of ginger rot.

Land Preparation

Land preparation starts with the receipt of early summer showers. The land should be ploughed 4-5 times or dug thoroughly to bring the soil to fine tilth. At the time of land preparation Thimat 10 G should be mix @ 10-15 kg/ha to check the attack of Nematodes.

Bed Preparation: Ginger should be grown in raised beds to avoid the water stagnation. Beds of about 50 cm width, convenient length and 15 cm heights are prepared at a spacing of 50 cm from each other. In areas prone to rhizome rot disease and nematode infestations, solarization of beds for 40 days using transparent polythene sheets is recommended.

Seed rate and Planting

Ginger is asexually propagated by portions of rhizomes known as seed rhizomes. Ginger plant does not produce true seeds. Carefully preserved seed rhizomes are cut into small pieces of 2.5-5.0 cm length weighing 20-25 g each having one or two good buds. 1500-2500 kg/ha is required to sowing. Prefer seed rhizomes weighing 20-30 g each and having at least 1 or 2 bud. The planted seed rhizomes are broken into pieces to ensure each piece has 2-4 sprouts. However in Sikkim whole rhizomes are planted without breaking to increase the recovery of mother rhizomes.

Seed treatment is done before sowing. The rhizomes are treated with 3% Diathane M 45 for 30 minutes. Then they shade dried for three-four hours or hot water treatment is done by dipping the rhizones in 50°C water for 10 mintus. Then they are planted in rows at 20 × 25 cm distance and 4-5 cm depth. The seed rhizome bits are placed in shallow pits prepared with a hand hoe and covered with well rotten farm yard manure and a thin layer of soil and leveled.

Manuring and Fertilization

Well decomposed FYM @ 25-30 t/ha has to be applied either on beds prior to planting or on pits at the time of planting. Soil amendments by application of neem cake @ 2t/ha helps in reducing rhizome rot and nematodes. In zinc deficient soils basal application of zinc fertilizer up to 6 kg zinc/ha (30 kg of zinc sulphate/ ha) gives good yield. The recommended dose of fertilizer is 75 kg/ha N, 50 kg/ ha P_2O_5 and 25 kg/ha K_2O. The fertilizers are applied in split doses as given in Table-1

Table 1: Manures and Fertilizers schedule in ginger per hectare

Fertilizers	Basal Application	40 days after planting	90 days after planting
FYM, t/ha	25-30	-	-
Neem cake, t/ha	2	-	-
N, kg/ha	-	37.5	37.5
P_2O_5, kg/ha	50	-	-
K_2O, kg/ha	25	-	-

Intercultural operations

Earthing up: Earthing up is essential to prevent exposure of rhizomes and provide sufficient soil volume for free development of rhizomes. It is done at 40 and 90 days after planting immediately after weeding and application of fertilizers.

Mulching : Mulching is a very important operation in case of ginger. Immediately after planting beds are covered with mulches consisting of green leaves/organic wastes up to 8-10cm thickness. It is essential to prevent soil splashing and erosion of soil due to heavy rain. It also adds organic matter to the soil, checks weed emergence and conserves moisture during the latter part of the cropping season. First mulching at the time of sowing @ 10-12 t/ha while second mulching is given at 40th day and 90th day with 5 tonnes of green leaves per ha should be done. The material used as mulch includes *Dhaincha* leaves, *Lantana camara* leaves, *Vitex negundo* etc. These mulches may reduce the infestation of shoot borer. After every mulching, cow dung slurry or liquid manure may be poured for enhancing microbial activity and nutrient availability.

Advantages of Mulching

a) It enhances germination.

b) Avoid splashing of soil particles.

c) Conserve soil moisture.

d) Source of organic matter.

e) Smothers weed growth and improve physical properties.

Crop Rotation and Mixed Cropping

It is an exhaustive crop so it is not desirable to grow ginger as same crop continuously. Hence, rotation with other crops is desirable. In Kerala, crop is rotated with tapioca, red gram, castor and chilli. In irrigated conditions it is rotated with paddy, Ragi, Ground nut, Maize, and Vegetables. Ginger is also known as or grown as intercrop in coconut, arecanut, coffee, orange, mango, guava plantation.

Harvesting

Ginger attains harvest maturity in 8-9 months hence, peak harvest starts from November and extend till January.

For green ginger or vegetable ginger :Harvesting is done from 6th month onwards i.e. before the rhizomes attain fibre maturity and is used in pickles, candy preparation and household uses. The rhizomes are thoroughly washed in water 2 to 3 times to remove the soil and dirt and sun dried for a day. On an average 5-8t/ha of rhizome is harvested.

For the purpose of Dry ginger : Crop is harvested between 8-9 months when leaves turn yellow and start gradually drying up. The clumps are lifted carefully with a spade and removed carefully adhered soil. For preparing dry ginger, the produce is kept soaked in water overnight. The rhizomes are then rubbed well to clean them. After cleaning the rhizomes are removed from the water and outer skin is removed with a bamboo splinters or wooden knife having pointed ends. The peeled rhizomes are washed and dried in sun uniformly for one week. The dry rhizomes are rubbed together in order to get rid of the last bit of the skin or dirt are called unbleached ginger.

Bleached ginger: To get good appearance, peeled scrapped ginger is dipped in the slurry of 2% slaked lime [$Ca(OH_2)$] water for six hours and then dried.

Yield

Fresh yield of ginger is 15-20 t/ha. The yield of dry ginger is 16-25% of the fresh ginger depending upon the variety, location etc.

Preservation of Seed rhizomes

Seed ginger has to be stored for about 4 or 5 months from the time of harvesting to planting. Keeping the seed rhizomes in pits is found to be the best method of storage. Healthy rhizomes are selected at the time of harvesting and are treated with mancozeb 0.3% or metalaryl mancozeb 0.125% for 30 minute as a prophylactic measure against soft rot. The treated rhizomes are then dried in shade and stored in pits dug in a cool place and protected from sun and rain. The pits should not be more than 1 meter deep. A layer of sand or sawdust is put at the bottom of the pit before storage. Put 2 cm thick layer of sand/ saw dust. Sufficient gap is to be left at the top of the pits for adequate aeration. Seed rhizomes in pits need inspection in twenty days to remove shriveled and disease affected rhizomes. Seed rhizomes can also be stored in pits dug in the ground under the shade of a tree provided there is no chance for water to enter the pits. In some areas, the rhizomes are loosely heaped over a layer of sand or paddy husk and covered with dry leaves in a thatched shed.

Plant protection

Insect-pests

1. **Shoot borer (*Conogethes punctiferalis*)**- It is the major pest infesting ginger. It appears in the month of July- October period. Larvae bore into the pseudo stems and feed on growing shoots. Infested shoots become yellow and drying occurs. The pest population is higher in the field during September-October

Control

i. Regular field surveillance and adoption of phytosanitary measures are necessary for pest management.

ii. Spray neem oil (0.5%) at fortnightly intervals found necessary.

iii. Light traps will be useful in attracting and collecting the adult moths.

iv. Spraying 0.1 per cent malathion.

2. **Rhizome scale (*Aspidiella hartii*)** – It infests rhizomes in the field (at later stages) and in storage. It causes withering in the plant.

Control

i. Treating the seeds with 0.075 per cent quinalphos for 20-30 minutes before planting may help prevent this infestation.

Diseases

Soft rot or Rhizome Rot (*Pythium aphanidermatum)* - It is a major disease of ginger. In this disease, collar region exhibits rotting and it spreads to rhizome and root. Leaves exhibit yellowing Symptoms.

Control

i. Soil solarisation should be done at the time of bed preparation

ii. Apply trichoderma at the time of planting.

iii. Provide adequate drainage.

iv Select disease free rhizome.

v. Treat seed rhizomes with 3% DM-45 for 30 minutes before storage and planting

vi. Drench the bed with 3% DM-45, add neem cake 2t/ha as basal dressing.

Processing of Ginger

Before processing ginger, it is recommended that a market survey is carried out. This will include information on the availability of raw material, availability of processing materials and equipment, access to markets and demand for the different ginger products.

Ginger based processed products

1. Dehydrated Ginger
2. Ginger Powder

Secondary and derived product

1. Ginger Paste
2. Ginger Bar/candies
3. Ginger Syrup
4. Ginger essential oils and oleoresins

Steps in Processing of Ginger

Selection of ginger rhizome

a) The stage of maturity of the rhizome at harvest determines its end use. At 8-9 months of age rhizomes are most suitable for drying.

b) When the rhizomes are harvested they should be handled with care to prevent injury. They should be washed immediately after harvest to obtain a pale colour. The wet rhizomes should not be allowed to lie too long in heaps as they are liable to ferment.

c) Care should be taken while removing the outer cork skin. It is essential to remove the skin to reduce the fibre content, but if the peeling is too thick, it may reduce the content of volatile oil which is contained near the surface.

d) During drying, the rhizomes loose about 60-70% of their weight and achieve a moisture content of 7-12%. Hence, care should be taken to prevent the growth of mould.

e) Dried ginger should be stored in a dry place to prevent the growth of mould. Storage for a long time results in the loss of flavour and pungency.

f) The appearance of the final product is very important especially when it is for export (not so important if the product is to be ground or used for oil extraction).

g) Content of volatile oil and fibre - especially for extraction of oils.

h) Level of pungency - especially for the extraction of oils.

i) Aroma and flavour - especially for the extraction of oils.

Preparation of ginger rhizomes

Selected rhizomes are washed under high pressure running water or in a washing tank. Rhizomes are killed by immersion in boiling water for 10 min. This also inactivates enzymatic processes, it is then sun-dried. Another method is to scrape, peel, or slice rhizomes prior to drying. Peeling or scraping is advised for reducing drying time, thus minimizing mould growth and fermentation. However, while this process decreases the fiber content by removing the outside corky skin, it also tends to remove some of the oils constituents, as they are more concentrated in the peel, and therefore reduces some of the pungency. The peeled rhizomes may be bleached to improve appearance.

Preparation of dried ginger

The most important factor is the cultivar of ginger grown. This determines the flavour, aroma, pungency and levels of essential oil and fibre. Fresh rhizome is harvested at an age of 8 and 9 months for drying. Cut the rhizomes manually or using a vegetable cutting machine into slices of 2-3 mm thickness. Dry the slices under sun or in a ventilated oven at 55°C for 18-19 hours or until brittle (6-7% moisture content). Store the dehydrated slices in High Density Polythene (HDPE) bags or Laminated Low Density Polythene (LDPE) packages. It has a shelf life of 12 months.

Preparation of ginger powder

The dehydrated ginger is ground into a fine powder which is then dried further at 80°C for about 30 minutes, after which it must be immediately sealed into HDPE or LDPE to minimize loss of aroma and avoid humidity. It has a shelf life of 12 months.

Preparation of ginger paste

After selection and preparation of ginger rhizomes, cut the ginger into pieces and blend into a thick paste. Ensure that a clean and pre disinfected blender is used (rinse blender before use with hot water). Add a little vegetable oil to the paste (6 % v/w i.e 60 ml for 1 kg of paste). Cook under low heat for 5 minutes, stirring continuously to prevent sticking of the paste on the sides of the pot. Fill in the paste in the pre sterilised jars, remove air bubbles with the help of a spatula or teaspoon, leaving a headspace of 5-10 mm. Add a teaspoon of warm oil on the surface of the paste before sealing. Pasteurise at 85-90ºC in a pasteuriser or in a pan of boiling water (the jars are placed in the pan when temperature of water is at 70ºC) for 10 minutes. Cool, label and store in a cool and dry place away from direct sunlight. It has a shelf life of 10 months

Preparation of ginger bar

Blend the ginger pieces (add around 150 ml of boiled and cooled water for each kg of ginger to ease blending) in a pre sterilized blender or juice processor. Filter the mixture through a muslin and collect the filtrate. Dilute the filtrate with boiled and cooled water in a 1:1 or 1: 2 ratio (depending on the pungency of the filtrate and also the desired pungency in the end product) and measure the volume of diluted filtrate. Add sugar to the filtrate at the rate of 600 g in 1000 ml filtrate. Cook under medium to low heat until filtrate starts thickening. For 1 litre of filtrate , mix 4% w/v of pectin i.e 40 g in 1000 ml filtrate with 100 g of sugar. Gradually, add the mixture of pectin and sugar to the filtrate; mix properly and cook under medium to low heat. The minimum Brix of 65° is required. Cook until mixture thickens and sets. Transfer the paste in a greased stainless steel plates or mould plates. Place the plates in a ventilated oven at 60ºC for 22-24 hours until the paste sets. Cut the paste into square pieces of 2.5cm × 2.5 cm. Coat the ginger bar with icing sugar. Label and store in a cool dry place.

Preparation of ginger syrup

Take 1 kg peeled ginger, add around 500 ml of water and 75g lemon peel and blend till a mixture is obtained. Filter the mixture in a muslin or cheese cloth to collect the ginger juice. Measure the volume of filtrate (the residue can be sun dried). Add 65% of sugar to the filtrate (650 g of sugar for 1000 ml of filtrate), warm the juice and monitor Brix. When Brix reaches 45°, add lemon juice (100-125 ml for each litre of the filtrate) and heat until final Brix of 50-55° is reached. Fill in disinfected HDPE bottles or sterilized glass bottles, when hot then seal bottles with disinfected covers or lids. Inverse the bottles for 5 minutes to allow the hot liquid to be in contact with the caps. Allow to cool, label and store in a dry place away from light. The syrup is very strong as ginger is very pungent (high in oleoresins) and dilution ratio with water prior to drinking is around 1:5 (according to taste).

Varieties

Suprabha	It is a clonal selection from Kanduli local. Fresh mean yield is 22.6t/ha with bright grey skin, it matures in 200 days, fiber 4.4%, oleoresin-6.7-8.9% and essential oil % is 1.8.
Suruchi	It is a Clonal selection. Fresh mean yield is 11.6 t/ha, Rhizome green yellowish, it matures in 218 days, oleoresin 10% and essential oil % is 1.9.
Surabhi	It is an X-ray induce mutant of a local cultivar. Colour – Dark graze, oleoresin – 10.2%, Yield potential is 4 t/ha and fiber content is 4%.
Surari	Fresh mean yield is 17.5 t/ha, it matures in 225 days, oleoresin 10.2% and essential oil % is 2.1.
Himgiri	Fresh mean yield is 13.5 t/ha, it matures in 230 days, oleoresin 4.3% and essential oil % is 1.6.
IISR Mahima	Released through selection, high yielding mature in 200 days, average yield 23.2 t/ha (fresh) with dry recovery of 23%, essential oil 1.72%, oleoresin 4.8% and fibre content 3.26%. Rhizomes are bold and the variety is resistant to root knot nematode.
IISR Rejatha	It is released through selection. It is a medium duration bold variety, with an average yield of 22.4 t/ha (fresh) with dry recovery of 19%, essential oil 2.36%, oleoresin 6.34% and fibre content 4.0%.
Rio-de- Janeiro	It recorded highest yield of 25-35 t/ha. Fiber content ranges from 5-19%.

References

Sasikumar, B., Saji, K.V., Antony, A., George, J.K.,Zachariah, T.J. and Eapen, S.J. (2003) IISR Mahima and IISR Rejatha- two high quality ginger (*Zingiber officinale* Rosc.) varieties. *Journal of Spices and Aromatic Crops* 12 (1): 34-37.

2.2 TURMERIC

Botanical Name	*Curcuma longa*
Family	Zingiberaceae
Origin	South East Asia, China
Chromosome number	2n=32, 86
Commercial value	Rhizome
Chemical constituents	Rhizome contains active curcumin compound while essential oil contains curcumin (Terpene)
Research Institute	Indian Institute of Spice Research, Calicut, Kerala.

Importance

Turmeric is an erect perennial plant grown as an annual crop for its rhizome (underground rootlike stem bearing roots and shoots). As a dried rhizome of a herbaceous plant, turmeric is closely related to ginger. The spice is also called "Indian saffron" due to its yellow colour. Curcuminoids are responsible for the yellow colour of turmeric. The underground rhizome imparts a distinctive flavour to food but it is also used to provide food with a deep, indelible orange colour. It is sold in the form of fine, dried, yellow powder. Turmeric is used in a wide variety of foods of the cuisines of Southern Asia but locally it also applies as an antiseptic for skin abrasions and cuts, purifier, liver tonic, *etc*. Curcumin (Terpene) is helpful in dissolving the cholesterol.

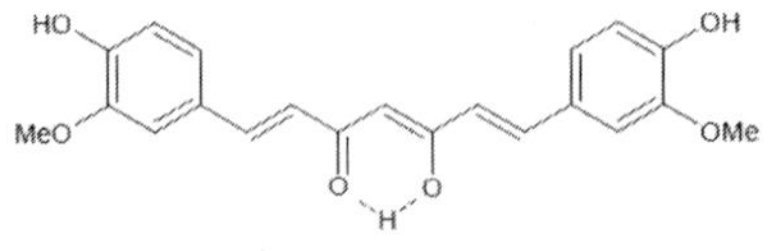

Curcumin

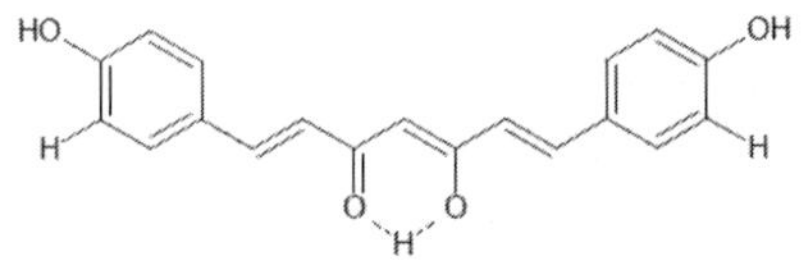

Bisdemethoxycurcumin

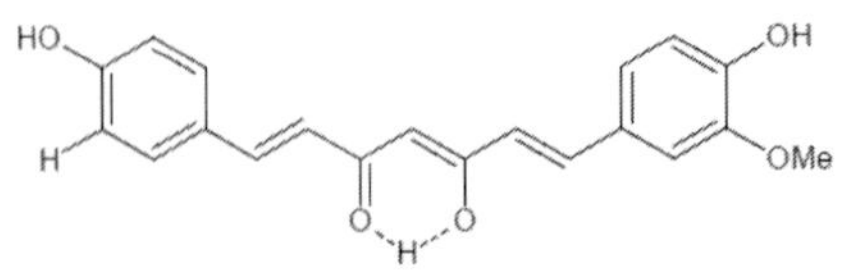

Demethoxycurcumin

Distribution

The plant originated in the Indian sub-continent and today India is the world's leading producer and consumer of turmeric. It is also produced in China, Taiwan, Bangladesh, Indonesia, Sri Lanka, Australia, Africa, Peru and The West Indies. In India the major turmeric growing states are Andhra Pradesh, Orissa, Tamil Nadu, Assam, West Bengal and Maharashtra.

Description of plant

It is an herbaceous perennial with a thick underground rhizome giving rise to primary and secondary rhizomes called gingers. The leaves are lanceolate with long leaf stalks. The flowers are borne on separate peduncle arising directly from the rhizome.

Soil

Turmeric can thrive in different kinds of soils such as sandy loam to clay loam or alluvial soils are suitable. Well-drained loamy soils are the best. Black red sandy loam soil rich in humus are good for turmeric. Soil should be rich in organic matter and uniform in texture. It cannot withstand water stagnation or alkalinity.

Climate

It grows in a wide range of climatic conditions, but does require rainfall of between 1000 and 2000 mm a year. Warm and humid climate, with 20-30° C temperature are required for successful cultivation. Growth ceases when, temperature falls below 20° C, and hence early-planted turmeric gives good yield. It can grow in locations that are up to 1220 m above sea level.

Cultivation

Sowing time

In western coastal regions of India where the rainfall is sufficient, early crop can be planted during April-May with the receipt of pre-monsoon showers. In Andhra Pradesh and Tamil Nadu planting is done in May-June or July-August.

Land Preparation

Land should be prepared through 3-4 ploughing in receipt of one shower of rains.

Seed bed preparation

There are basically two systems of planting (1) Raised beds and (2) Ridges and furrows but line sowing can also be done. Width of seed bed should be 1 m and length as convenient which may be up to 3m. 50cm Spacing is maintained between beds while, plant to plant Spacing is kept 15- 20 × 20-25cm. In case of raised beds it is raised upto a height of 15 cm.

Propagating material

Seed of turmeric consists of rhizomes. Turmeric emerges from mother rhizome is called fingers. Both mother and finger rhizomes are used. The fingers are cut into pieces, each 4-5 cm long with 1-2 buds. Mother rhizomes are planted as such or split into two, each having one or two sound bud. Large sized and healthy mother rhizomes are preferred since they give 50% more yields than the finger rhizome and also give good growth. Whole or split mother rhizomes weighing 35-44 g are used for planting. Rhizomes should be healthy and disease free. Rhizomes are treated with 3 % Dithane M-45 and 5% Malathion for 30 min before storing.

Seed Rate

It varies according to type of planting material, spacing and weight of rhizomes.

Mother rhizomes : 2000-2500 kg/ha.

Finger rhizomes : 1500-2000 kg/ha.

As an intercrop in fruit garden : 400 - 500 kg/ha.

Manuring and Fertilization

Table 2: Manures and fertilizers – In irrigated areas

Fertilizers	Basal Application	30 days after planting	60 days after planting	90 days after planting	120 days after planting
FYM, t/ha					
Neem cake, kg/ha	200	-	-	-	-
N, kg/ha	25	25	25	25	25
P_2O_5, kg/ha	60	-	-	-	-
K_2O, kg/ha	60	-	-	-	-
$FeSO_4$	30	-	-	-	-

Intercultural operation

Mulching : The crop is to be mulched immediately after planting with green leaves or banana pseudo stems or sugar cane trash-(@ 12-15 ton per ha. It may be repeated for 2nd time after 50 days with the same quantity of green leaves after weeding and application of fertilizers.

Weeding : Weeding should be done in 60,120,150 days after planting depending upon weed intensity. It can be grown as a mixed crop with chillies, Colocasia, Onion, Brinjal, Maize, Ragi etc.

Curing of Turmeric

Quality and proportion of cured and dry produce to green produce depends on the variety. The mother rhizomes have higher curing percentage than 'fingers.

Harvesting

Turmeric is harvested when the plants are between 7 and 10 months of age, when the stems and leaves start to dry out and die back. It is harvested from January to March. Early varieties mature in 7-8 months. Medium in 8-9 months. The whole plant is removed from the ground, taking care not to cut or bruise the rhizomes.

Sweating

The leaves are removed from the plant and the roots carefully washed to remove soil. Any leaf scales and long roots are trimmed off. The side (lateral) branches (which are known as the fingers) of the rhizomes are removed from the main central bulb (known as the mother). The mothers and fingers are heaped separately, covered in leaves and left to sweat for one day. The 'mothers' are the preferred material for planting the following year.

Yield

20-25 tonnes per ha fresh rhizomes can be obtained.

Plant protection

Pest

Shoot borer- *Conogethes punctiferalis*. Larvae bore into the pseudo stems and feed on the growing shoot resulting in yellowing and drying of infested shoots.

Control – Spray Malathion @ 0.1% at monthly interval from July to October.

Diseases

Rhizome and root rot – complete drying of leaves, rhizomes get decomposed and decayed.

Control: Drench the soil with 0.1% wet cerasan.

Leaf blotch – C.O.-*Taphrina maculans* Small oval, rectangular, irregular brown spot on either side of leaves.

Control: Spray of Dithane M-45 @ 0.2%.

Root knot Nematodes (*Meloidogyne and Radopholus Similes*).The infested rhizomes have water soaked areas in the outer tissues and burrowing nematode (*Radopholus similes*) are the two imporant nematodes causing damage in turmeric. Leaves dry from tip, root system.

Control: Nematode control is basically centered around prevention because once a plant is parasitized it is extermely difficult to eradicate the nematode without destroying the host plant.

Use biocontrol agent *Verticillium chlamydosporium* and VAM.

Processing of Turmeric

It is valued for its yellow pigment curcumin (diferuloylmethane) which varies from 4-8% in the dried rhizomes. Curcumin in turn contains Curcumin I (almost 94%), Curcumin II (6%) and Curcumin III (0.3%).

Processing of turmeric

Turmeric based processed products:

1. Dehydrated or dried turmeric
2. Turmeric Powder

Secondary and derived product

1. Curry powder
2. Turmeric oleoresins
3. Turmeric oils

A) Dehydrated or dried rhizome

Turmeric is mostly imported as a whole rhizome, which is then processed into powder or oleoresin.

1. Cleaning

a) It is done by removing soil adhering to rhizomes and fingers.

2. Boiling

a) Cleaned fresh rhizomes are kept in tin pan for boiling two to three days after harvesting.

b) While boiling, the water level should be 5-7 cm above the rhizomes

c) For quick and proper boiling, the rhizomes are covered with turmeric leaves 5-7 cm in thickness and plastered with mud and cow dung.

d) Due to use of cow dung, fingers get artificially coloured with lead chromate (chemichrome) which is poisonous. To avoid this CFTRI, Mysore has proposed to boil rhizomes in lime water (20 g Na_2CO_3 + 20 ml HC1 per 50 kg rhizomes).

e) Boiling is continued for about 45 min. to 1.5 hour.

f) The rhizomes are boiled in water until froth appears at the surface and the typical turmeric aroma is released.

g) However, if not boiled for long enough, the rhizome will be brittle. The optimum stage is reached when the rhizomes are soft to touch or can be pierced by a blunt piece of wood. Over cooking spoils the colour and fingers become light and get broken during polishing.

h) Because the fingers and bulbs are of different size and thickness to each other, it is important that they are cured in different batches. The curing time is dependent on the age, variety and size and thickness of the rhizomes and varies from one to four or sometimes six hours. It is also affected by the size of the batch smaller batches generally produce a higher quality dried turmeric as the curing process is easier to control and monitor.

3. Drying

a) The cooked fingers or bulbs are dried until they have a final moisture content of 5-10%.

b) Traditionally the rhizome pieces are laid on clean concrete floors and dried in the sun. This method takes 10 to 15 days, depending on the climate and the size of the rhizome pieces. It is important that the rhizome pieces are not placed in direct sunlight as this will cause the colour to fade. Using a mechanical drier will result in a better colour and a higher quality product. These include the tray drier, cross flow air tunnels, solar driers and cabinet driers. The optimum drying temperature is 60° C. Temperatures higher than this result in a darker coloured product.

c) The material is stirred 2-3 times to ensure uniform drying.

d) Improper drying results in hard and brittle rhizomes.

4. Sorting

Mother rhizomes and fingers if cooked together are separated before polishing.

5. Polishing

After drying the rhizomes are polished to remove the rough surface. Polishing is done to smoothed external surface of rhizome and for colour improvement.

Methods of Polishing

a) **Rubbing**: Rubbing against hard surface or trampling under feet after wrapping in gunny bags.

b) **Hand Polishing:** By hand or by shaking the rhizomes in a gunny bag filled with stones.

c) **Using Polishing Drum**:These are used in many places. They are very simple, power driven drums that have an abrasive metallic mesh lining having a capacity of 20-30 kg per rotation.

d) **Turmeric polishing machine**: This polishing machine is easy to operate. It is simple and compact so that it takes less space. It is run by electric motor but in case of power cut a pair of gear to run the machine has been provided. Machine polishing enables greater and furnished quality turmeric as well as reduces nutrient loss (Moghe *et al.*, 2012).

6. Grading

Quality specifications for turmeric are imposed by the importing country. They refer to the cleanliness of the product rather than on the eating quality. Bulk rhizomes are graded into fingers, bulbs and splits.

B. Turmeric Powder

Dried rhizomes are ground to approximately 60-80 mesh particle size. Since curcuminoids, the colour constituents of turmeric, deteriorate with light and to a lesser extent, under heat and oxidative conditions, it is important that ground turmeric be packed in a UV protective packaging and appropriately stored. It is a major ingredient in curry powders and pastes.

Varieties

Important varieties are broadly classified into three groups depending on duration as follows-

i. Short duration 'Kasturi' types-It belongs to *Curcuma aromatica* maturing in seven months and having the acceptable odour for culinary purposes. Short duration types are good yielder of dried turmeric and are rich in volatile oil but possess low curcumin.

ii. Medium duration 'Kesari types-It belongs to *Curcuma longa* maturing in eight months and is referred to as intermediary or CI types. Medium duration types are generally good in curcumin content and volatile oils but relatively poor in curing.

iii. Long duration types-It belongs to *curcuma longa* maturing in nine months and are referred to as CII types. This type is moderately good for both rhizome yields and other quality constituents.

Krishna — This variety is high yielding (72 q/ha) and released for general cultivation in Maharashtra. It is developed by clonal selection from Tekurpeta. Crop duration is eight months.

Ca 66J — It is recommended for rainfed conditions because of its lower rhizome storage losses.

Co.1 — It is vegetative mutant selection derived by X-ray irradiation of 'Erode Local' turmeric variety at 5 kR. Bigger sized, mother as well as finger rhizomes obtained.

BSR-1 — Bigger sized mother rhizomes and medium sized fingers. It has higher curing percentage. It is induced mutant from Erode Local.

Swarna — Accession PCT-8 subsequently released as Swarna. It had maximum yield potential (31.2t/ha) and high curcumin content (8.7%).

Suguna — It is clonal selection from Assam type. It is short duration (190 days) crop.

Suvarna	It is a clonal selection from Singhat, Manipur. It is medium duration (210 days) crop.
Prabha	It is an open pollinated progeny selection and suitable for all over India. It is of 205 days duration with 37t/ha rhizome yield. Rhizomes are reddish yellow in colour, with 6.52% curcumin, 15.0% oleoresin and 6.5% essential oil.
Prathibha	It is an open pollinated progeny selection and suitable for all over India. It is of 225 days duration with 39.12 t/ha rhizome yield. Rhizomes are reddish yellow in colour, with 6.52% curcumin, 16.2% oleoresin and 6.2% essential oil.

2.3 CHILLI

Botanical Name	1. *Capsicum annum* L.(Fruits are long or small, but pungency is high). 2. *Capsicum frutescence* L. (Large and inflated variety with very little pungency).
Family	Solanaceae
Origin	South America
Chromosome number	2n=24
Commercial value	Red pungent dried paprika fruits or, more precisely, pericarp
Chemical constituents	Capsaicin: (E)-N-[(4-hydroxy-3-methoxyphenyl) methyl]-8-methyl-6- nonenamide Capsanthin: (3R,3'S,5'R)-3,3'-dihydroxy-beta, kappa-carotene-6-one Capsorubin: (3S,3'S,5R,5R')-3,3'-dihydroxy-kappa,kappa-carotene- 6,6'-dione

Importance

Capsicum frutescence is a perennial chilli with small sized pods which are highly pungent. It is commonly known as 'bird chilly' and 'Tabasco'. Chilli is one of the most valuable crops of India. The crop is grown largely for its fruits all over the India. It is used in India as a principle ingredient of various curries, and chutneys. It is also used for vegetables, spices, condiments, sauces and pickles. As a medicine it is used as an counter irritant in Lumbago, Neuralgia, and Rheumatic disorders.

Capsicum has a tonic and carminative action. Taken inordinately it may cause gastro-enteritis. The enzyme isolated from chilly is used in the treatment of certain type of cancers. Oleoresin capsicum is used in pain balms and vaporubs. Dehydrated green chilly is a good source of vitamin 'C'

Dry chillies are used for curry powder. Fruit and seeds of *Capsicum annuum* L contains the major flavouring and colouring principles of this spice; the major flavouring principle is capsaicin ($C_{18}H_{27}NO_3$); the major colouring principles are Capsanthin $C_{40}H_{56}O_3$) and (Capsorubin $C_{40}H_{56}O_4$) while the yellow-orange colour is from beta-carotene and violaxanthin. A wide variety of other coloured compounds are known to be present. Capsaicin, an alkaloid, is extracted from chillies and is used to medicine. Colour is expressed as ASTA units. Above 120 ASTA is desirable. Pungency of Chilli is expressed as Scoville Heat units. Pure Capsaicin rates over 15,000,000 Scoville Units.

Capsanthin, the major carotenoid in ripe fruits, contributes up to 60% of the total carotenoids. Capsanthin and capsorubin increase proportionally with advanced stages of ripeness; with capsanthin being the more stable of the two. Paprika and paprika oleoresin is currently used in a wide assortment of foods, drugs, and cosmetics, as well as for improving the feather colour of flamingoes in zoos.

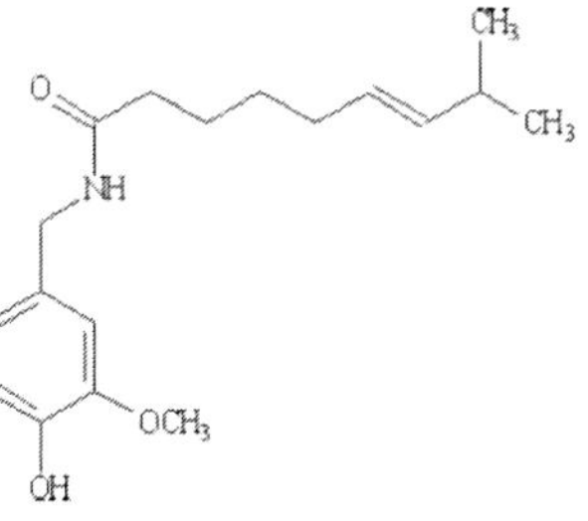

Capsaicin

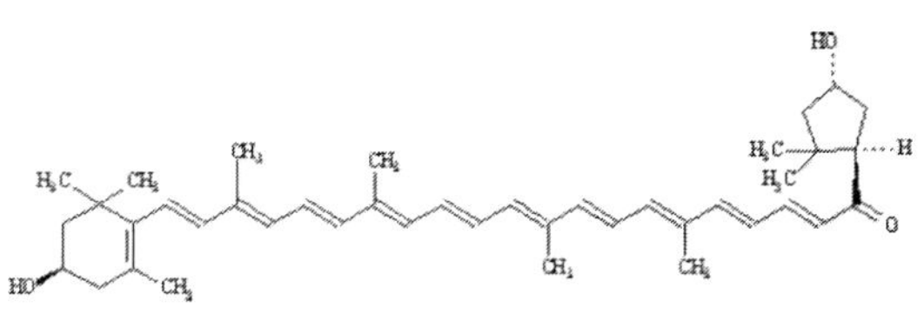

Capsanthin

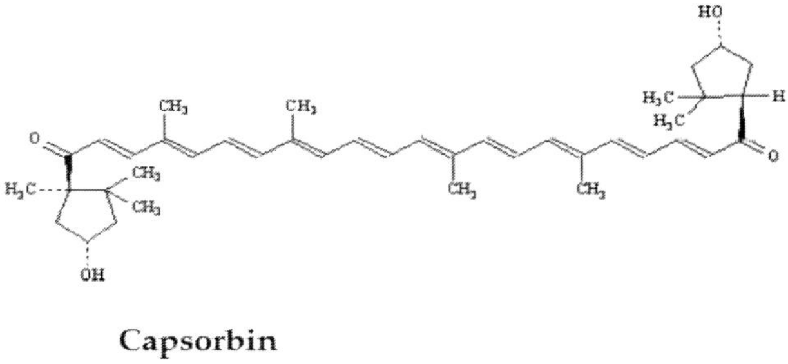

Capsorbin

Distribution

India is one of the leading producing country of the world. Andhra Pradesh, Maharashtra, Karnataka and Tamil Nadu are the major growing state in the country. The other states growing chilli commercially are Madhya Pradesh, Punjab and Bihar.

Description of plant

It is herbaceous or semiwoody annual or perennial. The leaves are ovate, tapering to a sharp point, entire upto 15cm length. The flowers are small, white and borne singly or in clusters of 2-3 in the axils of the leaves. The fruits are diverse in shape and it is variety based.

Soil

Chillies can be grown on a wide range of soils but for high yields, light loam soils are the best. In sandy soils, the crop can be grown with success provided adequate irrigation and fertilizers are available. The optimum soil pH required by the crop is 6.5-7.0. The field is prepared with deep ploughing followed by 2-3 cross harrowing, for drainage, and should be well levelled.

Climate

Chilli grows well in warm and cold climate. Ideal temperature ranges is 20-25° C. Low moisture in soil during blossom development and fruit formation causes the bud, deblossom and fruit drops. Excessive rainfall is detrimental to the crops, because it brings about defoliation and rotting of the plant. As a rainfed crop, it is grown in areas receiving an annual precipitation of 25-30 inches.

Cultivation

Seed rate and Time of sowing

Seed rate is about 1-1.5kg per ha. Before sowing seeds are treated with Cerasan or Agrosan as a plant protection measure against seed borne diseases. It can be grown as autumn crop in the month of June-July and as spring summer crop in the month of November.

Transplanting- Chilli seedlings are transplanted when they are about 4-6 weeks old. Extra transplanting in a second seed bed gives an earlier and higher yield. After transplanting, light irrigation should be given in order to establishment of the plant.

Method of planting and spacing

Ridges and furrow type is used. Seedlings are raised on raised bed. Spacing for rainfed crops is 60 × 45 cm and for irrigated crops is 60 × 60 cm.

Manure and fertilizers

At the time of field preparation 9 to 10 tonnes / hectare of FYM or compost is applied. In chilli for rainfed crop 50 kg N and 25 kg P/ha should be applied. Half dose of N along with full dose of P is applied at the time of transplanting. Remaining half dose of N is applied 30 days after transplanting. In case of irrigated crop 100 kg N, 50kg P and 50 kg K should be applied per hectare. P and K are applied as basal and N Fertilizers are applied in four equal doses. First dose is applied at the time transplanting remaining doses are applied at 4th, 11th and 13th week after transplanting.

Irrigation

About 7-8 irrigations are required for maintaining soil moisture and prevent deblossoming and fruit drop.

Interculture operations

2-3 shallow hoeing should be given to the soil to kill the weeds and provide soil mulch during early stages of growth- Application of weedicides for controlling the weeds is found effective. Lasso @ 1.5 litre per hectare with one hand weeding or Tok-E 25@ 2 litres per hectare with one hand weeding were effective in controlling the weeds.

Harvesting

Chillies which are used for vegetable purposes are generally harvested while they are still green but full grown. Chillies are harvested at red stage for canning purpose. Chillies used for drying are harvested at full ripe-stage.

Yield

The yield varies according to the system of cultivation. The yield of dry chillies of rain fed crop is 5 - 8 quintal/ha and that of irrigated crop is 15 - 25 quintal/kg per acre. The proportion of dry to fresh chilli varies from 25 to 40 per cent.

Drying of chilli

Chillies are perishable having 60-80 percent moisture content but for storage, the moisture content should be 10 per cent. In India the ripe chilli are dried under sun for 10-15 days depending upon weather condition. Excessive delay in drying results in growth of microflora and subsequent loss of quality. Commercially it is dried at 54° C for 2-3 days.

Plant Protection

Insect-pest

Chilli thrips: Thrips is the common pest which affects the crop throughout its life cycle. But they are more severe when plants begin to flower. These small insects suck the sap from the foliage and lacerate the leaf tissue, which result in curling of leaves and fall down of flowers prematurely.

Control: Thrips can effectively be controlled by spraying Dimethoate (Rogor 30 EC) or Monocrotophos (Monocil) 1 ml per litre of water at fortnightly interval.

Aphids: Aphid suck the sap from the plants; they generally attack the crop in winter months and at the later stages of the crop. The quality of the produce is spoiled by imparting blackish colour to the calyx and pods. They also serve as a vector to virus.

Control: Aphids can effectively be controlled by spraying the crop with Dimethoate (Rogor-30 EC) or Methyl Parathion (Metacid 50 EC) @ 1 ml (or Phosphamidon (Demicron-100 EC) @ 0.5 ml per litre of water.

Diseases

1. **Damping off-**It is a serious disease of chilli seedlings and mainly occurs in nursery. The disease infected seedlings rot at ground level and then the plants fall over ground. The seedlings die in patches.

Control

The seed bed should be treated with Formalin before sowing of seeds. The seeds should be treated with water (30 minutes at 52°C) or Cerasan or Agrosan G.N. before sowing of seed. The seedlings in the nursery should be sprayed with any fungicides at a regular interval.

2. **Anthracnose-**Dark sunken spots are formed of fruits and pink or dark coloured dots appear in the centre of the sunken spots. Due to this spots, the fruits rot and fall. The fungus may cause "Die back" of the twigs also. Die back disease attacks mainly the upper portion of the plants spreading gradually from the top to downward; as a result the branches dry up. Moist weather, shade and heavy dew favour the occurrence of the disease.

Control

Treatment of seed with Cerasan before sowing, removing and burning of attacked plants or branches and spraying the disease affected crops with Mancozeb (Dithane M-45) @ 2.5g per litre of water.

3. **Leaf curl**-The disease affected leaves becomes small in size accompanied by downward curling. The leaves may fall off in case of sever attack. The disease usually spreads through insect vectors such as thrips and aphids.

Control

Control of insect vectors by spraying the crop with Dimethoate (Rogor -30 EC) or Monocrotophos (Monocil) @ 1 ml per litre of water indirectly helps to check the spread of this disease.

Varieties

Jwala	This variety is highly pungent, light red in colour, short and the seeds are compact. Its harvesting season is September to December. The Capsaicine contents is 0.4%.
Pusa Jwala	Developed from 'NP46A' × 'Puri Red'. Pods are long (above 10 cm), thin and pendent, red in colour, contain low seeds and mildly pungent. Prone to viral diseases and root knot nematodes. Yield is 10-25 q/ha. Capsicin content of oleoresin is 8.0%. Adaptable to all over India.
Bhagya Lakshmi (G–4)	Released in 1975. Recommended for irrigated regions all over the country, pods are olive green, turning to dark red on ripening, calyx is deeply shaped, seed content 38-40%, fairly tolerant to pests & diseases, useful for green chilli also. Maximum yield is 50q/ha.
Andhra Jyothi (G-5)	Pods are short and stout with conical shape, Pericarp is thick, glossy with deep red colour, Seed content 46%. Maximum yield is 50q/ha.
Bhaskar (LCA 235)	Released in 1985. Early maturing, compact plant type, flowers with yellow anthers serve as marker gene, fruits thin 5-6 cm long, calyx deeply cup shaped, seed content - 45%, highly pungent, tolerant to thrips and mites, green chillies with less moisture content can withstand transport to gulf countries, tolerant to viral diseases to some extent.
Arka Lohit	Released by IIHR, Bangalore in the year 1990. A pure line selection from IIHR 324 (local collection). Plants tall. Fruits dark green, smooth, straight, turning deep red on maturity

	(Capsanthin 0.205%). Fruits highly pungent (Capsaicin 0.708%) suitable for irrigated and rainfed cultivation. Duration 180 days. Yield 25 t/ha (green chilli) and 3 t/ha (dry chilli).
Arka Suphal PMR 57	Indeterminate with a plant height of 80-90 cm. Dark green foliage fruits straight, smooth with pointed tip, 7-9 cm long, fruit colour green changing to deep red. Yield is 25t green and 3 t dry chilli/ ha. Resistant to powdery mildew and field tolerant to viruses.
Pusa Sadabahar	Pods erect 8 cm long with 6-14 per cluster. Ripe pods bright red, pod yield 15-20q/ha (dry). Resistant to CMV, TMV and leaf-curl viruses.
Kashi Anmol	Developed through selection and released during 2006. Plants are determinate, bushy with nodal pigmentation on stems. Fruit pendant, attractive green in colour; suitable for green fruits production under chilli-wheat/chilli potato cropping system. First picking takes 50 days after transplanting and crop duration 130 days.

Hybrid

Arka Harita: High yielding chilli F1 hybrid developed by using MS line. Plants tall (1m) and spreading (90cm). Fruits medium long (10 cm) with width 1 cm green fruit yield 30 t/ ha and dry fruit yield 6 t/ ha in 150-160 days. Fruits are dark green and turn red. Tolerant to powdery mildew and viruses.

Arka Shweta: High yielding chilli F1 hybrid developed by using MS line. Plants medium tall (95 cm) and spreading (82.5 cm). Fruits long (13.2 cm) with 1.3 cm width. Fruits are light green which turns red on maturity. Fresh fruit yield is 38.4 t/ ha and dry fruit yield of 6 t/ ha. Crop duration is 140-150 days.

Arka Meghana: F1 hybrid developed using male sterile line. High yielding with 30-35t / ha fresh yield and 5t/ ha dry yield. Fruits green, slightly wrinkled, long (10-12 × 1.2cm), pendent, turning deep red on maturity. Very early taking 60-65 days after transplanting for first harvest. Field tolerant to viruses and sucking pests. Fresh fruit yield 30-35tonnes/ha, dry fruit yield 4.5-5.0 tonnes/ha.

References

T. John Zachariah. Indian spices and Exploration of its intrinsic quality, Indian Institute of Spice Research, Calicut, Kerala, India.

2.4 ONION

Botanical Name	*Allium cepa* Linn
Family	Amaryllidaceae
Origin	Central Asia and Mediterranean region
Chromosome number	2n=16
Commercial value	Bulbs
Chemical constituents	Allyl propyl disulphide
Research Institute	Directorate of Onion and Garlic Research, Pune

Importance

It is one of the most important export oriented spices having nutritional as well as medicinal value. They are rich in sulphur, fibers, potassium, vitamin B, vitamin C and they are low in fat, cholesterol and sodium. The pungency in onion is due to sulphur-bearing compound in the volatile oil allyl propyl disulphide. It is used against sun stroke. Onions have anti-biotic, antiseptic, antimicrobial and carminative properties to help stay away from infections. It is an immediate cure for fever, common cold, cough, sore throat, allergies etc. A mixture of onion juice and honey can cure these problems easily. An onion a day can cure insomnia or sleeping disorders. This will surely give you a good night sleep. They can improve digestive system. If you have digestion problem, then onions can cure it by increasing the release of digestion juices.

Distribution

In India, onion is grown on large scale. The main growing states are Maharashtra, Gujarat, Karnataka, Tamil Nadu, Orissa, Madhya Pradesh, Uttar Pradesh, Andhra Pradesh, Bihar and Punjab.

Description of plants

It is a hardy bulbous plant. It is an annual for bulb production and biennial for seed production. The edible portion is modified stem which is known as bulb and develops underground. It has bisexual flowers and is highly cross-pollinated crop, which is affected by insects.

Cultivation

Climate

It thrives better in cool season and long days favours bulb formation. But the requirement of day length of different varieties may differ. *Kharif* onion varieties require day length of 10-11 hours, whereas *rabi* varieties require relatively higher temperature and 12-13 hours day length. Long day varieties do not bulb under short day whereas short day varieties, if planted under long day, develop bulbs. However, varietal response to climate is different. It germinates well in 20-25° C, vegetative growth requires mild climate and prefers 13-21° C and bulb formation requires 15-21° C and for ripening, dry weather is favourable. Drought conditions are very harmful for plant growth and bulb formation. Long days are beneficial in the formation of bulbs. Besides photoperiod, light intensity and quality, temperature, planting time, plant growth, availability of nutrients, cultivars, presence of growth substances and management practices affects the formation of growth.

Soil

Soil for onion should be deep friable and highly fertile. Onions can, however, be grown in all types of soils. Sandy soil needs more and frequent irrigation and favour early maturity, whereas heavy soils give rise to mis-shapen bulbs causing problem in digging of bulbs. It prefers light soils rich in humus, friable, well drained with ability to retain soil moisture for the good size development of bulb. Avoid heavy soils for onion cultivation. The optimum range of pH should be 5.8 to 6.5 except Hissar-2 and Punjab Selection which are salt tolerant varieties.

Land preparation

The field should be ploughed to a fine tilth by 4 -5 times with a sufficient interval between two ploughings. The ploughings should be done by tractor-drawn implements or *desi* plough. Ploughing should be shallow as most of the onion roots penetrate to a depth of not more than 5-6 cm. The planking should be done for proper leveling. The field is then divided into beds and channels. The normal width of a bed should be about 1.8 m.The length may, however, vary according to level of land but it is better to have small-sized beds for proper irrigation and other cultural operations.

Seed rate- In order to raise seedlings for planting of one hectare about 8-10kg healthy seeds will be sufficient. When it is planted through bulb or bulbils about 1000 to 1200kg are needed for one hectare.

Raising of seedlings- Onion seeds are sown in nursery beds to raise seedlings. Best time for sowing early-*kharif* crop is February-April in South India, May-June in other parts, August-September for late-*kharif* and October-November for *rabi* in plains all over the country. In hills (higher altitudes), March-April is ideal sowing time. It is raised on the raised nursery beds and the total area required for raising seedlings for one hectare planting of onion crop is about 500sqm area i.e. about twentieth part of a hectare prior to sowing. Raised beds should be of size 3 m×0.6 meter×20-30 cm. About 70cm distance is kept between 2 beds to carry out operations of watering, weeding etc. The surface of beds should be smooth and well-levelled. Raised beds are necessary to avoid problem of waterlogging in heavy soils. In sandy soils, however, sowing can be taken up in flat beds. Sowing should be done in lines spaced at 5-7 cm distance. Seeds are treated with Thiram 2-3g/kg of seeds before sowing. Application of *Trichoderma viride* @ 1250 g/ha is also recommended to manage damping off and raise healthy seedlings.

Sowing time

The seeds are sown in the month of September to November for *rabi* season crop and May-June for *kharif* season crop.

Transplanting

Seedlings are transplanted when they are about 10-15cm in height and 6-8weeks old in the beds of convenient length and width at the distance of 15 to 20cm between rows and 10cm between plants and 2 to 3 cm deep. If lower aged seedlings are transplanted establishment is poor and if over-aged seedlings are planted there is a problem of premature bolting. Prior to transplanting of seedlings, the upper most portion of the seedlings to the extent of 20-25% is cut in order to facilitate quick establishment of seedlings. Removal of tops reduce the rate of transpiration rate because the volume of foliage is reduced.

Manure and fertilizers

In general, soil for onion growing should be liberally manured and fertilized. The requirement of nutrients depends on soil type, region of growing, varieties and removal of major nutrients. About 20-25 tonnes of farmyard/ha should be applied one month before transplanting / planting or sowing and mixed well in the soil. Whole quantity of phosphorus, potassium and half of nitrogen should be mixed in the soil before transplanting. Remaining half of nitrogen should be given as topdressing in single dose in bulblet method of planting in big onions and bulblet method of planting in multiplier onion 20-25 days after planting. In transplanting method, topdressing is taken up in 2 split doses and first dose should be applied at 30 days after transplanting whereas second dose at 45 days after transplanting. The topdressing must be completed before initiation of bulbing. Delayed application results in thick neck or doubles.

Irrigation

Just after transplanting of seedlings light irrigation is essential. The quantity of water and its frequency of its application depends on growth, soil type and planting season. The critical stages for irrigation in onion are bulb formation and enlargement stages.

Interculture and weed control

Onion is a shallow rooted crop. Shallow hoeing once or twice will helpful in plant growth and keeping down the weeds to get a good yield of onion. Stamp @ 3.5 litres/ha applied 3days after transplanting plus 1 hand weeding at 45 days in *rabi* onion. While in *kharif* onion Stomp @ 2.5 lit/ha with 1 hand-weeding is very effective. Oxyflurofen @ 150-250 g/ha with one hand-weeding is also effective.

Harvesting

Harvesting takes in 45-90 days from field setting for green onions and 65-150 days for bulb depending upon variety / types. Bulbs are considered to be mature when the neck tissues begin to soften and tops are about to abscise and decolourize. Development of red pigment and characteristic pungency of variety are also important harvesting indices of onion.

Onion for sale as dried bulbs or for storage should be harvested after tops have started falling over. Best time to harvest *rabi* onion is one week after 50% tops have fallen over. In *kharif* season since tops do not fall, soon after the colour of leaves changes to slightly yellow and tops start drying, red pigmentation on bulbs develop and also true shape and size develop, bulbs are harvested and kept in windrow for drying the tops. Small picking and multiplier onions are harvested when 50% - 75% of tops have fallen over. The leaves are cut leaving about 2-2.5 cm tops above the bulb after complete drying. Leaving the tops intact until complete drying increases dry-matter content which might be due to greater loss of water from the bulbs with intact foliage or due to movement of materials from the tops to the bulbs. If tops are cut too close, the neck does not close well and decay organisms have easy access to the bulb. Early harvesting results in sprouting of bulbs while late harvesting results in formation of secondary roots during storage. In *kharif* season late harvesting results in doubles and bolting.

Yield

The average yield of onion bulb is 300-400quintals in *rabi* season and 200-250quintals in *kharif* season. The seed yield is 6-8 quintals per hectare in case of bulb to seed method and 8-10quintals in case of seed to seed method.

Plant Protection

Diseases

Damping off- Causal organism- *Fusarium oxysporum* fungus. The disease is more prevalent in Northern and Eastern parts of the country during *kharif* season, causing 60-75% damage. Two types of symptoms are observed.

- **Pre-emergence damping off :** The fungus kills the radicle and plumule of seed before emergence from soil.
- **Post-emergence damping off :** The pathogen attacks the collar region of seedlings on the surface of soil. The collar portion rots and ultimately the seedlings collapse and die.

Control

1. For managing the disease, healthy seeds should be selected for sowing. The seed should be treated with thiram @ 2g/kg before sowing.
2. Continuous raising of nursery in a particular plot should be avoided. Top soil of nursery should be treated with thiram @ 5 g/m^2 area and nursery should be drenched with the same chemical @ 2 g/litre of water at fortnightly interval.
3. Soil solarization using 250-gauge polythene for 10 days before sowing and application of *Trichoderma viride* in soil @ 1,250 g/ha is also effective to control damping off to a considerable extent.

Purple blotch- It is caused by fungus *Alternaria porri*. The disease occurs under favourable conditions of temperature 28-30°C and 80-90% relative humidity. It is more common in *kharif* season. Small, sunken, whitish flecks with purple coloured centres are common symptoms occurring on leaves and flower stalks. Further, large purple area develops forming dead patches. The intensity of disease varies from season-to-season.

Control

1. Use healthy seeds.
2. Follow crop rotation.
3. Spray mancozeb @ 0.25% or chlorothalonil @ 0.2% should be sprayed at fortnightly interval commencing from one month after transplanting. The sticker triton/sandovit should be mixed in spray solution.

Downy mildew- Its causal organism is *Perenospora destructor.* Symptoms on the surface of leaves or flower stalks violet growth of fungus is noticed which later becomes pale-green yellow and finally leaves or seed stalks collapse.

Control

For managing the disease effectively, onion bulbs meant for seed crop should be exposed to sun for 12 days or heated for 4 hours at 41° C to destroy the fungus. Spraying of 0.2% zineb also gives good control.

Onion smut- It is caused by *Urocystis cepulae*. The fungus survives in soil in area where temperature remains below 30°C. Since the fungus remains in soil, disease appears on cotyledons of young plants soon after it emerges. Smut appears as elongated dark, slightly thickened areas at the base of seedlings. The black lesions appear near the base of scales on planting. The affected leaves bend downwards abnormally. On older plants numerous raised blisters occur near the base of leaves. The lesions on plant at all stages often expose a black powdery mass of spores. The fungus becomes inactive at soil temperature of 26-27°C and above.

Control

1. Treating the seeds with captan or thiram @ 2.5 g /kg of seed before sowing.
2. Seed-bed soil treatment with methyl bromide at 1 kg/25 m^2 or alternative application of formaldehyde solution @ 8 ml/litre of water /10m^2 is effective in controlling the disease.

Stemphyllium blight

It is caused by *Stemphyllium vasicurium* fungus Yellow/orange spots are seen on leaves and immediately the size increased & ultimately leaves become dry.

Control

30DAT, apply the fungicide at the interval of 10-15 days. Mancozeb @ 0.25% or 0.1% Propiconazol should be used as a fungicide.

Pests

Onion thrips-Thrips (*Thrips tabaci*) is most injurious insect pests of onion. Onions infected with thrips develop spotted appearance on leaves which turn pale-white blotches due to drainage of sap. Sometimes 50-60% losses in bulb crop are observed due to thrips attack. Seed production including viability of seeds are hampered due to attack of thrips. The infested plants show stunted growth with twisted leaves. If the infestation comes in early stage of growth, bulb formation stops completely and plants die slowly. Bulbs in store are also attacked by thrips.

Control

1. Malathion (0.1%) or metasystox (0.1%) and cypermethrin (0.01%) or deltamethrin (20 ml/ha) should be sprayed along with triton or sandovit to control the insect.
2. Soil application of carbofuran granules @ 1 kg ai/ha is also recommended.

Onion maggot-The adults of onion maggot (*Hylemia antiqua*) appears like house fly. The flies lay their eggs on old leaves or on soil and larva enters into soil and damage disc portion of onion bulb. Infested plants turn yellowish brown and finally dry up. The affected bulbs rot in storage as infestation leads to secondary infection by pathogenic organisms.

Control

1. Crop rotation should be followed.
2. Thimet should be applied in the soil before transplanting.

Varieties

Bhima Super	A red onion variety from DOGR has been identified for release for *kharif* season in Chhattisgarh, Delhi, Gujarat, Haryana, Karnataka, Madhya Pradesh, Maharashtra, Odisha, Punjab, Rajasthan and Tamil Nadu. It can also be grown in late *kharif*. It is reported to have an average yield of 20 - 22 t/ha in *kharif* and

	40 - 45 t/ha in late *kharif.* Bulbs attain maturity within 100-105 days after transplanting (DAT) in *kharif* and 110 -120 DAT in late *kharif.* It produces mostly single centered bulbs.
Bhima Dark Red	This onion variety from DOGR has been identified for release in Chhattisgarh, Delhi, Gujarat, Haryana, Karnataka, Madhya Pradesh, Maharashtra, Odisha, Punjab, Rajasthan and Tamil Nadu for *kharif* season. Average marketable yield is 20-22 t/ha. It is recommended particularly due to its attractive dark red flat globe bulbs. It attains maturity within 95-100 DAT
Bhima Red	This variety already recommended for *rabi* season in Maharashtra and Madhya Pradesh, is also recommended for *kharif* season in Delhi, Gujarat, Haryana, Karnataka, Maharashtra, Punjab, Rajasthan and Tamil Nadu. It can also be grown in late *kharif.* Maturity is 105-110 DAT during *kharif* and 110-120 DAT during late *kharif* and *rabi* seasons. The average marketable yield in *kharif* season is 19-21 t/ha, in late *kharif* season is 48-52 t/ha and it is 30-32 t/ha in *rabi* season. It can be stored up to 3 months in *rabi.*
Bhima Shweta	This white onion variety already recommended for *rabi*, is also recommended for release for *kharif* in Chhattisgarh, Gujarat, Karnataka, Madhya Pradesh, Maharashtra, Odisha, Rajasthan and Tamil Nadu. TSS is around 11-12° Brix and it matures within 110-120 DAT. It has medium keeping quality and can be stored up to 3 months. Average marketable yield during *kharif* season is 18 - 20 t/ha and in *rabi* is 26-30 t/ha.
Bhima Shubhra	This white onion variety has been recommended for Chhattisgarh, Gujarat, Karnataka, Madhya Pradesh, Maharashtra, Odisha, Rajasthan and Tamil Nadu for *kharif* season. It is also recommended for late *kharif* in Maharashtra. It matures in 110-115 DAT during *kharif* and 120-130 DAT in late *kharif.* TSS is 10-12° Brix. It is a medium storer with a capacity to tolerate environmental fluctuation. Average marketable yield during *kharif* is 18 - 20 t/ha and during late *kharif*

	36-42 t/ha.
N-53	Recommended by the Department of Agriculture, Maharashtra during 1990 and also released at the National level by All India Coordinated Vegetable Improvement Project in 1987. Suitable for *kharif* season in western and northern India. Duration is 90-100 days. Bulbs are medium, flatish round and red in colour. Average yield is 250 quintals per hectare.
Pusa Red	Developed from IARI, New Delhi. Bulb medium in size, average weight 70-90g, bronze red in colour, flat to globular shaped and less pungent. Keeping quality is good. Plants mature in 140-145days after transplanting. TSS is about 12-13%. Average yield is 25 t/ha. It can be grown both in late *kharif* and *rabi* season in Maharashtra. Suitable for export purpose.
Agrifound Dark Red	Bulbs are dark red, globular in shape, 46cm in size with tight skin, moderately pungent. TSS is 12-13%. Plant matures in 95-110 days after transplanting. Yield is 30-40 t/ha. Average keeping quality. Recommended for kharif season. Suitable for export purpose.
Agrifound Light Red	The bulbs are of globular shape with tight skin, light red colour and 46cm in size. TSS is 13%. It has good keeping quality. Plants mature in 160-165days after sowing. Average yield is 30-32.5t/ha. Recommended for growing during *rabi* all over country. It can be grown in late *kharif* season.

Hybrid

Arka Akshay	It is a tri-parental synthetic variety with dark red globe shaped bulbs. Average bulb weight is 115 g. Yield 45 t/ha in 130 days.
Arka Bheem	It is a tri-parental synthetic variety with red to pinkish-red elongated globe shaped bulbs. Average bulb weight is 120 g. Yield 47t/ha in 130 days.
Arka Lalima	Developed from cross between MS 48 x Sel 14-1-1 and released during 1985. Attractive deep red bulbs, uniform, globose and firm. Stores well for 4-5 months.

	Can be grown in *kharif* and *rabi*. TSS 10-12%. Maturity 140 DAT. Yield 45 – 50 t/ha. Identified for southern dry zone of Karnataka state.
Arka Kirtimaan	Developed from cross between MS 65 x Sel 13-1-1 and released during 1986. Attractive red bulbs, uniform, globose and firm. Stores well for 4-5 months. Can be grown in *kharif* and rabi. TSS 8 – 10%. Maturity 130 DAT. Yield potential 40 – 45 t/ha. It is also identified for southern dry zone of Karnataka.

2.5 GARLIC

Botanical Name	*Allium sativum* L.
Family	Alliaceae
Origin	Europe
Chromosome number	2n = 16,24,26
Commercial value	Bulbs
Chemical constituents	Diallyl disulfide (60%), diallyl trisulfide (20%), allyl propyl disulfide (6%), a small quantity of diethyl disulfide and probably diallyle polysulfide
Research Institute	Directorate of Onion and Garlic Research, Pune

Importance

The potency of garlic has been acknowledged for more than 500 years. In the ancient times, garlic was used as a remedy for intestinal disorders, flatulence, worms, respiratory infections, skin diseases, wounds, symptoms of aging, and many other ailments. Through the middle ages into World War II, the use of garlic to treat wounds surfaced repeatedly. It was ground up or sliced and was applied directly to wounds to inhibit the spread of infections. Garlic contains more than 200 chemical compounds. Some of its more important ones include: volatile oil with sulphur-containing compounds: (allicin, alliin, and ajoene), and enzymes: (allinase, peroxidase and myrosinase). Allicin is what gives garlic its antibiotic properties and is responsible for its strong odour. Ajoene contributes to the anticoagulant action of garlic. Garlic also contains citral, geraniol, linalool, Aphellandrene and B phellandrene. Vitamin and Mineral Content in garlic includes vitamins B especially B-1, vitamin C, vitamin A, flavonoids, ascorbic acid, phosphorous, potassium, sulphur, selenium, calcium, magnesium, germanium, sodium, iron, manganese and trace iodine. Seventeen amino acids are found in garlic, including eight essential ones (Anonymous, 1995).

S-allyl cysteine (SAC) is one of the water-soluble organo-sulfur compounds in garlic that can be detected in the plasma, liver, and kidney after oral intake. Its concentration increases during extraction and aging. From several studies, SAC has proven to be a stable, odourless, water-soluble compound with the ability to lower cholesterol, serve as an antioxidant, inhibit the cancer process, and protect the liver from toxins (Roy *et al.*, 2009). It has also exerted some effect on glucose tolerance for both hypo-and hyperglycemia by reducing insulin requirements to control blood sugar.

Garlic powder is mass-produced as a flavoring agent for condiments and processed foods. Garlic cloves are sliced or crushed, dried and pulverized into powder. Powdered garlic can be taken in capsules and can be purchased in deodorized form. Garlic capsules are a convenient way to supplement the diet with garlic and are good for heart disease, high blood pressure and to fight infections of any kind. Garlic juice can be taken for digestive disorders, infectious diseases and for atherosclerosis (hardening of the arteries).

Distribution

Garlic is a native of West Asia and Mediterranean area. China, Korea, India, USA, Spain, Argentina and Egypt are the major garlic growing countries.

Description of plant

Garlic plants have 6-12 flat, narrow "regular" leaves and can reach from just under 10 inches to over 6 feet in height There are two basic types of garlic: hardneck and softneck. Hardneck garlics are characterized by hard, woody central stalks that extend down to the basal plate at the bottom of the bulb Bulbils are generally smaller than cloves but, like cloves,can vary in size and number. Soft neck garlic's are thought to have evolved from hardneck garlic's. Softnecks have a non- evolved from hardneck garlics .Softnecks have a non-woody pseudostem formed from overlapping leaf sheaths and rarely send up a flower stalk, unless stressed by climatic. Garlic as a bulb made up of cloves. Garlic bulbs can range in size from 1.5 to 3 inches in diameter depending on variety and cultivar and can have from 4-60 cloves of various shapes and sizes (25, 46,

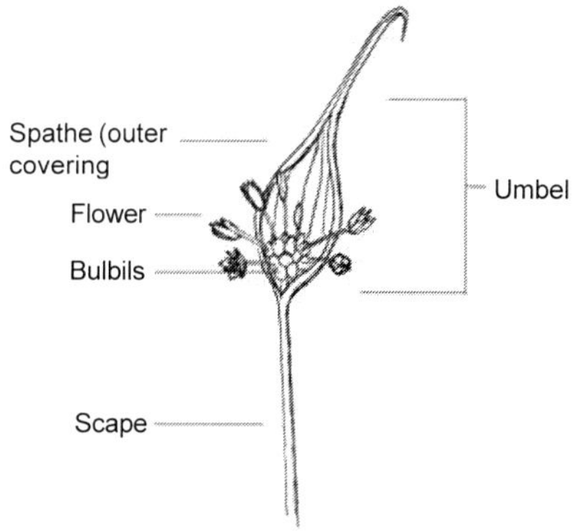

(*Source*: Garlic Herb Society of America)

70). Cloves are enclosed in a white or pinkpurple tinged papery membrane and are actually swollen "specialized leaves".

Climate

Garlic is a frost tolerant crop. It requires cool and moist climate during growth period and warm dry weather during maturity.

Soil

Garlic can be cultivated in various types of soils, but well drained fertile loamy soil is suitable for this crop.

Land Preparation

Land should be brought to fine tilth by shallow ploughing 4-5 times, giving sufficient interval between two ploughings. Land should be levelled and divided into small plots and channels to facilitate proper irrigation and intercultural operations.

Time of Planting

The planting differs from region to region. In Gujarat, planting of Garlic is done during October-November. In Maharashtra, M.P., Karnataka and A.P., it is planted from August to October and in northern part of India from September to November.

Spacing

General recommendation of spacing for Garlic is 15cm row to row and 10 cm from plant to plant.

Manures and Fertilizers

General recommendation of manures and fertilizers for garlic in normal soil condition is 50 mt FYM, 100 kg Nitrogen, 50 kg Phosphorous and 50 kg Potash per ha. Full dose of FYM should be incorporated in the soil at the time of land preparation. Half dose of Nitrogen, full dose of Phosphorus and Potash may be applied as basal dose at the time of planting and remaining Nitrogen is applied after one month of planting.

Intercultural operation and weeding

Generally 2 to 3 weeding are recommended. First weeding should be done one month after planting, second one month after first weeding and third weeding at later stage of crop. Hoeing should be done at the time of bulb formation ie. 45 days after planting. It helps in setting of bigger well filled bulbs.

Irrigation

Garlic being a shallow rooted crop, it needs frequent irrigation. There should be no moisture scarcity during growth period. Generally, irrigation should be given at an interval of 8-15 days depending on soil and climate. When crop reaches its maturity (plants pseudo stem begin to bend /break or dry) irrigation should be stopped.

Harvesting and yield

The crop becomes ready for harvest when the tops turn yellowish or brownish and show sign of drying. It takes about 4-5 months after planting to reach maturity. After harvesting, bulbs are dried in shade. Bulbs should be stored in cool dry place having proper aeration. Yield obtained is 10-20 t/ha.

Plant Protection

Pests

Stem & Bulb Nematode (*Ditylenchus dipsaci*): It is a tiny worm, which damages the tissues of garlic and reduces the yield significantly. Basal portion of the mature plants swell and become spongy, the leaves and stems become rotten.

Control: Soil fumigation with Dichloropropene-Dichloropropane (D-D) mixture before planting @ 500 kg per ha is recommended.

Onion Thrips (*Thrips tabaci*): This is a small insect which hides in the leaf axils and feeds on leaves turning it to white or silvery in colour.

Control: Spray of 0.2% solution of Carbosulfan or 0.1% Fipronyl insecticide is effective in controlling thrips.

Mites (*Rhizoglyphus callae & Glycohaqus domesticus*): These insects attack bulbs stored for longer periods.

Control: Spray Dicofol@0.2% or 0.05% Sulfur is recommended for control.

Diseases

i) **White Rot (*Sclerotium cepivorum*):** This is also called as Allium root rot. Yellowing and wilting of leaves are the major symptoms of attack of this fungus. It causes rotting of bulbs in later stage.

Control: It can be controlled by spraying of 0.15% solution of Carbendazim or 0.25 % solution of Dithane M-45 at the time of planting in rows.

ii) **Penicillium Rot (*Penicillium corymbiferum*):** Decay caused by penicillium may occur in field and storage both. This fungus generally invades thick fleshy part of the cloves in storage and softens it, later shrivels and finally blue-green spores are seen. Very young plants are affected by this fungus in the field.

iii) **Stemphyllium Blight (*Stemphyllium vesicarium*):** Small yellow – orange streaks on the leaves. Soon they develop into elongated spots and turn grey at centre, finally it develop conidiophores and conidia of the fungus. These spots coalesce and blighted portion can be seen which gradually covers entire leaf.

Control: Spraying of 0.25% solution of Mancozeb at regular interval of 15 days after appearance of first symptoms, can be used for control of the disease.

iv) **Purple Blotch (*Alternaria porri*):** Symptoms of this disease appear on leaves. Small whitish sunken lesions/ purple blotch are seen on the leaves, which enlarge rapidly. The leaves fall gradually.

Control: Spraying of 0.25% solution of Mancozeb at regular interval of 15 days after appearance of first symptoms, can be used for control of the disease.

v) **Cercospora Leaf Blight (*Cercospora duddiae*):** The symptoms of this disease also appear on leaves. Small irregular dark brown spots scattered on lamina are seen in the plant affected by this disease. These spots gradually coalesce and foliage becomes blighted.

Control: Spraying of 0.2% solution of Ziram or Captan or 0.3% solution of Copper oxychloride at fortnightly intervals should be done to control this disease.

vi) **Powdery Mildew (*Leveillula taurica*):** Distinct pale yellow patches of variable size coupled with powdery mass on the lower side of leaves are main symptoms of this disease.

Control: Spraying of 0.2% solution of Sulphur fungicides should be done to control this disease and it should be repeated at regular intervals of 15 days after appearance of the symptoms.

vii) **Mosaic:** Garlic plants affected with mosaic virus show typical chlorotic mottling and stripes on the first emerging leaf followed by pale yellow broken stripes on matured leaves. As the virus is transmitted by aphids, controlling the vector is useful.

Control: Spraying of 0.05% solution of Monocrotophos is recommended to control mosaic.

Physiological Disorders

Some physiological disorders like sprouting of bulbs, are observed in some varieties of Garlic. This disorder is not of permanent nature and varies from variety to variety.

Varieties

Yamuna Safed (G-1)	The variety was notified in the year 1991 by Govt. of India. It was developed by mass selection from a local collection obtained from Delhi (Azadpur) market. The bulbs are compact, silvery white skin with creamy flesh. Diameter 4.0 cm to 4.5 cm. Sickle shaped cloves, 25-30 in number, size index of bulb 12-15 cm^2, diameter of cloves 0.8 to 1.00 cm. The variety is tolerant to insect pests and diseases like purple blotch, stemphylium blight and onion thrips. TSS 38%, dry matter 39.5% and good storer. Yield 150-175 q/ha. It is recommended for cultivation all over the country.
Yamuna Safed-2 (G-50)	The variety was notified by Govt. of India in the year 1996. It was developed by mass selection from a local collection obtained from Karnal area in Haryana. The bulbs are compact attractive white creamy flesh, bulb diameter 3.5-4.0 cm, size index 11-12 cm^2, number of cloves 35-40, diameter 0.75 - 14 cm. Clove size index 1.75-2.5 cm^2, 10 bulb weight 160-240 g, TSS 38-40%, dry matter 40-41%. Average yield 150-200 q/ha. The variety is recommended for Northern India.
Yamuna Safed-3 (G-282)	The variety has done very well in Northern parts and also in Central parts of India. It was developed by mass selection technique from a local collection obtained from Dindigul (Tamil Nadu) in 1990. The leaves are wider than other varieties.

	Bulbs are creamy white and bigger sized (5-6cm dia). Size index 27-29 cm^2, diameter of cloves 1.2-1.5 cm. 15-16 number of cloves per bulb, TSS 38.42%, dry matter 39-43%, medium storer. Average yield 175-200 q/ha. The variety is suitable for export.
Yamuna Safed-4 (G-323)	This variety was notified by Government of India in 2006. It was developed by mass selection technique from a local collection obtained from Jaunpur, Uttar Pradesh in 1988. Leaves are wide, Bulb- compact, attractive creamy white colour, creamy flesh, bulb diameter 4-5 cm, size index 14-16 cm^2, no. of cloves 18-23 per bulb, diameter of cloves 0.75-1.0 cm, cloves size index 1.8-2.0 cm^2, matures in 165-175 days. Yield 200-250 q/ha. Storage quality is better than Yamuna Safed-3. Suitable for exports.
Agrifound White	The variety was notified by Govt. of India in 1989. The variety was developed by mass selection from a local collection obtained from Biharsharif area in Bihar. The bulbs are compact, silvery white with creamy flesh. Bigger elongated cloves with 20-25 in numbers. Diameter 3.5 to 4.5 cm size index 12-15 cm^2. The variety is susceptible to purple blotch and stemphylium blight which are common in the northern parts.The TSS is 41%, dry matter 42.78% and is a good storer. The yield is 130 q/ha. It is recommended for cultivation in the areas where there is not much problem of purple blotch or stemphylium blight in *rabi* season.
Agrifound Parvati	This variety was developed in 1992 by selection from an exotic collection obtained from Hong Kong market. The variety is long day type and as such is suitable for cultivation in mid and high hill of Northern states. Bulbs are of bigger size (5-6.5 cm), creamy white colour with pinkish tinge. Size index 16-72 cm^2, diameter of cloves 1.5 to 1.8 cm, 10-16 cloves in number, tolerant to common disease. Average yield 175-225 q/ha, medium storer. Suitable for export.

References

Roy, H., Lundy, S. and Kalicki, B. (2009). Health Benefits of Garlic. Pennington Nutrition Series, Number 20.

Anonymous, 1995. *Garlic* -Nature's Amazing Nutritional and Medicinal Wonder Food. Woodland Publishing, Inc.pp 5-6.

2.6 CORIANDER

Botanical Name	*Coriander sativum* L.
Family	Apiaceae/Umbelliferae
Origin	Mediterranean region
Chromosome number	2n=22
Commercial value	Leaves and seeds
Chemical constituents	Dextro-linalol
Research Institute	Indian Institute of Seed Spices, Rajasthan

Importance

Coriander a perennial herb is used for garnishing, making chutneys or in breads. Health benefits of coriander are numerous. Its dark green colour makes it more attractive and appealing to eat. Coriander has its unique taste. It is also a rich source of vitamins (Vit B and C) and minerals. It also contains Vit K that prevents Alzheimer. Regular use of coriander helps to improve the overall health and debars several deficiency diseases. It help in fighting against Salmonella bacteria. In addition it is used as carminative. It finds its medicinal use in preventing digestive disorders. Coriander seeds consumed in the form of tea helps in kidney ailment, mouth ulcers, swelling, and lessens blood cholesterol.

Distribution

Main growing states in India are Rajasthan, Gujarat, Tamil Nadu, Andhra Pradesh and Madhya Pradesh.

Description of plant

It is andromonoecious, highly cross pollinated in nature, it is a tropical crop and can be grown throughout the year (except very hot season) for leaf purpose but for higher grain yield, it has to be grown in specific season. It is an annual herb and mainly cultivated for leaves and seeds. Fruits have fragrant odours and pleasant aromatic taste. The odours and taste are due to the essential oil content which varies from 0.1-1% in dry seeds. It masks the offensive odours in pharmaceuticals preparation. It has medicinal property too. Fruits are said to have carminative, diuretic, tonic, and aphrodisiac property.

Soil

All types of soils are sufficient for this crop. Black cotton soils with high retention of moisture are best under rain fed conditions.

Climate

It requires dry and cold weather free from frost especially during flowering and fruit setting. Cloudy weather during flowering and fruiting stage favours pest and disease incidences.

Cultivation

Sowing time

In the northern and central part of India it is mostly grown as *Rabi* season crop and hence sowing is done between middle of October and middle at November still late sowing is recommended in place, vulnerable for frost damage. Under rainfed conditions, it is sown during September and October at the onset of North East monsoon and harvested during January – February.

Field preparation

For raising a rainfed crop, the land is ploughed 3-4 times following rains and field must be planned immediately to break the clod and to avoid soil moisture. For irrigated crop the land is ploughed twice or thrice and bed and channels are formed.

Seed rate and Method of sowing

About 10 – 15 kg seeds per hectare is used for sowing. Seed stored for 15-30 days record better and early germination than freshly harvested seeds. Seeds soaked in water for 12-24 hours before sowing also enhance better germination. The seeds are split into two halves by rubbing and treated with Thiram at 2g per kg of seeds. For irrigated crop, sowing is generally done in rows spaced at 30-40 cm. with 15 cm hills. Soil depth up to 30 cm 3-5 seeds are sown in a hill and later on thinned to two plants per hill. Under rainfed conditions seeds are broadcast and covered with soil. Germination takes place in 10 to 15 days.

Manure and Fertilizer

In irrigated condition 10 tonnes/ha FYM should be applied at the time of last field preparation, 30 kg/ha Nitrogen, 40 kg/ha Phosphorous and 20 kg/ha Potash should be applied for proper growth of coriander.

Plant protection

Grain mould – It is caused by *Helminthosparium* sp. *Alternaria* sp., *Carvularia* sp., and *Fusarium* sp.

Control: It can be controlled by spraying Carbendazim @ 0.1% applied at 20days after fruit set.

Powdery mildew- *Erysiphe polygoni.*

Control: It can be minimized by spraying of Wettable sulphur (0.25%) or 0.2% solution of Karathane twice at 10-15 days interval.

Varieties

CO - 1	Tall plants, many umbels suitable for green and grains.
Gwalior No.5365	It is suitable for Madhya Pradesh, resist to fusarium wilt, medium seed size, and 110 days duration.
Moroccan	High oil content, medium seeded, duration 140-160 days and seed yield is 10-12q/ha.
CO - 2	Essential oil content of seed is high and varies from 0.30-0.48 percent.
CO - 3	It is reselection from Accession 695. It has high yield potential of 275.6kg/ha in *kharif* and 644kg/ha in *Rabi* season.
Swathi	It is developed through mass selection, duration- 80-85 days. Average dry yield 850kg/ha, early maturing type, escapes powdery mildew incidence due to early maturity.
Rajendra Swati	A mass selection from germplasm type. It is resistant to stem gall disease. It matures in 110 days, yield 1200-1400kg/ha.

2.7 FENUGREEK/GREEK HAY

Botanical Name	Trigonella foenum graceum (Common methi) Trigonella corniculata (Kasuri methi)
Family	Fabaceae
Origin	It has two areas of origin: the Indian sub-continent and the Eastern Mediterranean Region
Chromosome number	2n=16
Commercial value	Seeds and leaves
Chemical constituents	Saponins, coumarin, fenugreekine, nicotinic acid, phytic acid, scopoletin and trigonelline
Research Institute	National Research Centre on Seed Spices, Rajasthan

Importance

It is cultivated annually herb reaching height of about 4-5 metre. It is used as a leafy vegetable, spices, medicinal plant. The dried seeds are aromatic and bitter, which have been used traditionally in India, China, Egypt and in some parts of Europe for its well known carminative, galactogouge, antibacterial, anti-inflammatory, insulinotropic, and rejuvenating effects. Although fenugreek is a legume, it doesn't always fix nitrogen. For this to happen, the right bacteria (*Rhizobium meliloti*) need to be present in the soil. These are more likely to be present, if fenugreek has been grown on the site before. Seeds are used as condiment, also used for dye making and extraction of alkaloids or steroids. The seeds are also reputed to have many medicinal properties including relieving joint pain, lowering blood sugar level and even restoring hair growth.

T. foenum-graecum leaf is used as a green, leafy vegetable and is a good source of calcium, iron, β-carotene and several vitamins. Its seeds are good source of protein (20-30%) high in tryptophan and lysine; free amino acids (4-hydroxyisoleucine, arginine, lysine, histidine); (25.8%), fat (6.53%), ash content (3.26%), crude fibre (6.28%), energy (394.46 Kcal/100 g seed) and moisture (11.76%). It contains lecithin, choline, minerals, B. Complex, iron, Phosphates, PABA (Para-Amino Benzoic Acid) and vitamins A and D. The important chemical constituents are saponins, coumarin, fenugreekine, nicotinic acid, phytic acid, scopoletin and trigonelline. In addition to these main components, some minor components like alkaloids (trigonolline, cholin, gentianine, carpaine etc), free unnatural amino acids (4-hydroxyisoleucine), and individual spirostanols and furastanols like diosgenin, gitogenin, yamogenin etc. It prevents constipation, removes in digestion, stimulates appetizing and diuretic.

The significance of *T. foenum-graecum* seeds is due to the defatted part, with high quality fibre including steroidal saponins and protein comparable to those of soybean. The seeds also have the alkaloid trigonelline with mucilage, tannic acid, yellow colour substance, fixed and volatile oils and a bitter extractive, diosgenin and gitogenin a trace of trigogenin.

Distribution

It is widely cultivated in warm temperate and tropical regions in the Mediterranean, Europe, and Asia. The major seed producing countries are India, Ethiopia, Egypt and Turkey. In India it is grown extensively in Rajasthan, Gujarat, Madhya Pradesh, Uttar Pradesh, Maharashtra and Punjab.

Description of Plant

Plant consider as an annual legume with white or yellow flowers. This robust herb has light green leaves and is 30-60 cm tall. The plant has a sharp, spicy aroma. It has a smooth hollow stem with alternate, single trifoliolate leaves borne on a short petiole with two small stipules at the base. Leaflets are oval, and slightly toothed. Flowers are yellow-white (white-flowered types) or sometimes purple-tinged (normal types) and develop in the leaf axils either singly or in pairs. It is a typical self pollinated crop. Plants flowers in 37 days after sowing and the duration of the flowering phase is 7-18 days. Pods are brown, slender, and sickle-shaped with a sharp beak at the end. They are approximately 7-15 cm long, each containing about 10-20 seeds. Pods mature within 60-90 days after sowing.

Soil

Clay loam soil is better for cultivation purpose. Optimum soil pH should be 6-7 for its better growth and development.

Climate

It has wide adaptability and can be successfully cultivated both in the tropics as well as temperate regions. Tolerant to frost and freezing weather. Its performance is good in low rain falls than in heavy rainfall areas.

Cultivation

Land preparation and sowing

Land is prepared by ploughing thrice and beds of uniform size are prepared. Broadcasting the seed in the bed and raking the surface to cover the seed is normally followed. But, line sowing is advantageous. Seeds are thinly dropped in shallow opened furrows of 20 to 25cm apart. Sowing is done during September to November, whereas, in hills sowing start from the month of March.

Seed rate

About 20-25 kg seeds are sufficient for one hectare area. Seeds germinate within 6-8 days depending on planting season and depth.

Manures and fertilizers

The recommended nutrient dosage for fenugreek includes 15 tonnes/ha FYM, 25 kg/ha Nitrogen, 25 kg/ha Phosphorous and 50kg/ha Potash. Nitrogen should be applied after each cutting for good leafy growth.

Irrigation

First irrigation should be given at the time of sowing and then provide at the interval of 6-7 days.

Inter cultural operations

Weeding, hoeing at the early stage of the crop growth is required to encourage proper growth. Thinning may be done on 20-30 days to keep the distance between the plants at 10-15 cm. and to retain 1-2 plants per hill.

Harvesting

In about 25-30 days shoots are nipped off 4-5 cm. above ground level and subsequent cutting of leaves may be taken after 15 days. It is advisable to take 1-2 cutting before the crop is allowed for flowering and fruiting when pods are dried. The plants are pulled out dried in the sun and seeds are there shed by beating with stick or by rubbing with hands.

Processing of fenugreek

Seeds are winnowed, cleaned, dried in sun. Stored in gunny bags lined with paper.

Processing of Fenugreek Gum Powder from Fenugreek Seeds.

The process of making fenugreek powder from the fenugreek seeds involves the following steps:

1. The seeds are placed in a heavy bottomed pan preferably one made from cast iron.
2. The seeds are then roasted over a low flame, while constantly stirring. Roasting reduces the bitterness of the seeds making it more palatable as well as adding flavour to it.
3. When the seeds are nicely browned remove from the heat and allow them to cool.
4. The seeds are then placed into a mixer which grinds the seeds to powder. Thereafter, a whizz is used to make the powder finer.
5. The powder is then stored in a tight jar and can be used for weeks on end.

Properties of Fenugreek Gum Powder

- Fenugreek gum powder is tasteless and odourless.
- The powder contains galactomannan which has the viscosity property which can turn milk into ice-cream, cocoa into chocolate and forms gum when dissolved in water.
- The gum is an emulsifier which is capable of mixing water and oil. This comes from the fact that fenugreek gum powder is a combination of galactose which has the hydrophilic property and mannan which is hydrophobic.

Yield

1200-1500 kg of seed and about 800-1000 kg of leaves per hectare can be obtained.

Plant protection

Root rot C.O. - *Rhizoctonia solani.* It is a serious disease of Methi.

Control – Drenching Carbendazim @ 0.05%. First at the onset of the disease and another after one month.

Varieties

C.S. 381	It is a high yielding variety and suitable for Madhya Pradesh. Seed yield is 18-20q/ha and mature in 150 to160 days.

Plume -55	It is also recommended for Madhya Pradesh. Seed yield is 17-18 q/ha and the crop mature in 140-150 days.
Plume-20	It is suitable for *methi* green leaf production. It gives 7-8 q/ha seed yield and 70-80 q/ha green leaves.
Co-1	It is a reselection from TG 2336. It is quick growing, and suited for intercropping.
Rajendra Kranti	It is a mass selection suited for pure as well as intercropping.
RMt-1	It is a pure line selection from Nagpur type. It is moderately tolerant to root rot and powdery mildew.
Prabha	It is developed through pure line selection. It is most suitable for Rajasthan and Gujarat. It is less susceptible to root rot and powdery mildew. Under favorable conditions it gives 15-28 q/ha yield.

2.8 BLACK PEPPER

Botanical Name	*Piper nigrum*
Family	*Piperaceae*
Origin	Tropical forests of Western Ghats of South India.
Chromosome number	2n=52
Commercial value	Seeds
Chemical constituents	Piperine
Research Institute	Indian Institute of Spice Research, Kerala

Importance

It is one of the important and earliest known spices produced and exported from India. Pepper oil contains a range of organic compounds, which belong to the group of terpenes and hydrocarbons. The major aroma compounds of black pepper are β-caryophyllene, sabinene, myrcene, limonene and linalool. Piperine ($C_{17}H_{19}NO_3$) is undoubtedly the major pungent principle in pepper, which probably comprises over 95% of the total pungent alkaloids present. In general, black pepper contains about 3-5% volatile oil and 8-16% oleoresin and 2-6% piperine.

Black pepper has been used for digestive disorders, such as indigestion, vomiting, diarrhea, and flatulence. It is also used to treat colds, coughs and sore throats. Black pepper is a common ingredient in *chai*, a spiced black tea beverage used throughout India that includes other herbs also believed to be "warming tonics," such as cardamom, cinnamon, cloves, ginger, and star anise. In Ayurveda, diabetes is believed to be a disorder of "*kapha* imbalance," due to "low digestive fire," and black pepper is used as a warming herb to help stimulate digestive processes. Black pepper is also used in Ayurveda to treat cholera, colic, headache, toothache, and urinary problems. Externally, it has been applied in paste form to treat skin diseases.

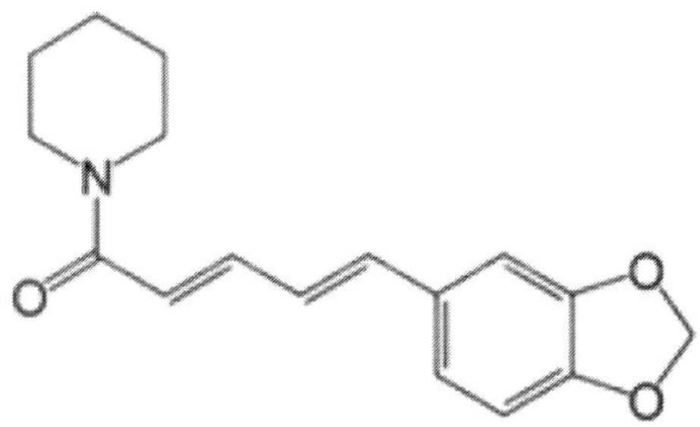

Piperine (1-[5-(1,3-Benzodioxol-5-yl)-1-oxo-2,4-pentadienyl]piperidine)

Distribution

Major countries involved in cultivation and trade of pepper are India, Indonesia, Sarawak (Malaysia), Brazil, Sri Lanka, Vietnam and China.

Description of Plant

It is a climbing evergreen plant and grows to a height of 10 m. or more. Five distinct type of stem can be identified in the shoot system of pepper vine. Main stem originates from a seed or from a stem cutting.

Runner shoots- These are produced from basal portion of the main stem, growing at right angle to the main stem. It is restricted up to 50 cm. from the ground.

Fruiting branches (Plagiotropes)- These are produced from the nodes of main stem and grow laterally more or less at 90° to main stem, bearing the spikes.

Top shoots (Orthotropes)-After a period of vertical growth, the top portion of the main shoots attain a bushy appearance with shorter, thicker inter nodes and profuse branching with large number of adventitious roots at the nodes. This portion of the main shoot is called top shoots or orthotropic.

Hanging shoots or Geotropes-In a fully grown vine, some of the plagiotropes at the top portion are seen to give rise to a special type of shoots, which hang down and grow geotropically.

Inflorescence- Catkin dioecious, hermaphrodile. Sex form – Monoecious. Majority of cultivated types of pepper are monoecious. Pepper is mostly dioeciously in wild form, but in the cultivated type the plants are most gynomonoecious or trimonoecious. Bisexual flowers are protogynoecious. Autogamy and geitonogamy type of pollination takes place in wild state. Cross pollination is essential for sexual reproduction while in cultivated pepper, mostly bisexual flowers are selected for genetic consistency and reliable fertility which is facilitated by self pollination.

Fruit- Its fruits are called berry. Dried ripe berries become black and wrinkled constituting black pepper. Black pepper yield both black and white pepper. Black pepper is made by drying ripe or unripe fruits under the sun; white pepper by soaking, treating and removing outer skin of berry before drying.

Cultivation

Climate

It is a plant of humid tropics requiring adequate rainfall and humidity. Crop tolerates temperature between 10° C and 40° C. It is best suited under humid climate with rainfall of 1500-2000 mm. It can grows well up to 1500 meters above mean sea level.

Soil

Black pepper grows in almost all types of soil. However, it thrives well in loose, well-drained soil with pH 4.5-6.0. Red laterite soils are more suitable.

Propagation

Black pepper is propagated by sowing seed, marcotting and by stem cutting. Of the three methods, stem cutting is generally employed using both terminal stem and lateral branches. The lateral shoots on rooting give rise to bush black pepper. Select planting materials from high yielding hermaphrodite mother plant. Divide stem into cuttings each with 3-5 internodes and cuttings are rooted in sandy seedbed under shade. Insert cutting at 450 angle 15 centimeters apart each way. When cutting has developed 4-7 new leaves, then it is ready for transplanting in the field.

In marcotting method, runner shoots from high yielding and healthy vines are kept coiled on wooden pegs fixed at the base of the vine to previous shoots from coming in contact with soil and striking roots. The runner shoots are separated from vines during February-March and after trimming leaves cutting of 2-3 nodes each one planted either in nursery beds or in polythene bags filled with fertile soil. If plastic sheet is available, cut into convenient pieces and use for enclosing soil or moss and tie both ends with string or vegetable tying material. In 3-4 weeks, each marcot can be separated from mother plant and set in the field. Adequate shade should be provided and watering be done frequently. The cuttings strike roots are become ready for planting in May-June. Rapid multiplication of black pepper has become popular in India. The main advantages of this method are: multiplication is rapid (1:40), better field establishment, well developed root system and more vigorous growth.

Site selection

When grown on sloppy land the slopes facing south should be avoided and the lower half of north and north eastern slopes preferred for planting.

Preparation of land and planting shade trees

With the receipt of first monsoon shower in the month of May-June, the primary stem cutting of *Erythrina indica* or *Grevilea robusta* are planted. The five standards are used at a spacing of 2.7 × 2.7 m accommodating 1100 vines/ha in monocropping system with the onset of monsoon, 2-3 rooted cuttings of pepper are planted individually in the pits on northern side of each standard. At least one node of the cutting should go below the soil for proper rooting. At a spacing of 2.5 × 2.5 m pits of $0.5m^3$ are dig. Cutting are planted nearly 30 cm away from the standard. In case of inter coiled root net close to trunk, pepper cuttings must be planted 100-120 cm away from tree trunk. After one year when the vines attained sufficient length it may be separated from temporary stake and lower leaves may nipped off. A narrow trench of 15cm deep and wide should be prepared. The vine may be placed in the trench in such a way that the growing tip is tied to the trunk while the other parts of vine are covered with the soil.

It starts to bear fruits three years after planting. Open space can be utilized for planting short-lived catch crops such as snapbeans, ginger, hot and sweet pepper at a distance of one meter away from black pepper rows.

Cultural practices

Black pepper needs weeding, mulching and pruning. When seeding reaches one or two feet high, nip off the top growing vine to induce more branching. Prune old and unproductive branches and crowding laterals. As the cutting grows, the shoots are tied to the standards as often as required. The young vines should be protected from hot sun during summer months by providing the artificial shade. Regulation of shade by looping the branches of standards is necessary not only for providing light to the vine but also for enabling the standards to grow straight. During the 2nd year, some cultural practices are repeated. However, looping of the standards should be done carefully from 1st year onwards.

Manure and fertilizer

For laterite soils which are low to medium in major nutrients, the fertilizer recommendation for pure pepper crops is 140 g. N. 55 g P_2O_5 and 270 g K_2O/ vine.

Table 3: Scheduling of fertilizers

Year	Rate	Time
1st year	only 1/3rd of the dose	May-June and Aug-Sep.
2nd year	2/3rd of the dose	May-June and Aug-Sep.
3rd year onwards	Full dose	May-June and Aug-Sep.

Method

Apply fertilizer at distance of about 30cm. all around the vine and covered with a thick layer of soil. Compost @ 10kg per vine during May and Lime @ 600g/ vine during April-May in alternate years should be given.

Harvesting

Pepper vines start yielding usually from the 3rd or 4th year. Vines start flowering in May-June.It takes 6-8 months from flowering to ripening stage. Harvesting start from November to January in plains and January to March in hills. Harvesting should be done when one or two berries on the spike turn bright to red, the whole spike is plucked. Berries are separated from the spikes by rubbing between the hands. After the separation, the berries are dried in sun for 7 days until the outer skin becomes black and shrunk or wrinkled appearance of commercial black pepper.

The white pepper of commerce is prepared from either freshly harvested barries or dried black pepper using special techniques such as water stepping (7 days) steaming (3-4days) decortications for removing dull white colour. The recovery of white pepper is about 25% of ripe berries while that of black pepper is about 33%.

Improved Varieties

Panniyar-1 This is the first released variety and is evolved through hybridization. It is a Hybrid between Uthirankotta × High Cheriyakaniakadan from Kerala Agril. University. The mean dry yield kg/ha is 1242. Piperine % is 5.3, Oleoresin% 11.8 and essential oil % is 3.5. It is not suited for heavily shaded area. The productivity in this variety is due to the recombination of favorable and complementary yield components from the diverse parents.

Panniyur-2 It is a Selection (Cul. 141) from open pollinated progeny of cultivar Balankotta. The mean dry yield kg/ha is 2570. Piperine % is 6.6, Oleoresin% 10.9 and no essential oil is recovered. It tolerates shade compared to other cultivar. It has non-pigmented growing tip.

Panniyur-3 It is a Hybrid between Uthirankotta × Cheriakaniakadan. The mean dry yield kg/ha is 1953. Piperine % is 5.2, Oleoresin% 12.7 and no essential oil is recovered. It has faint pigmentation on growing tip. It is a late maturing cultivar.

Panniyur -4 Selection from Kuthiravally Type. The mean dry yield kg/ha is 1277. Oleoresin% 9.2 and no Piperine and essential oil is recovered. It is a stable yielder.

Panniyur -5 Open pollinated progeny selection from Perumkodi. The mean dry yield kg/ha is 1098. Piperine % is 5.5, Oleoresin% 12.3 and essential oil % is 3.8. It is tolerant to shade.

Panniyur -6 Clonal selection from Karimunda. The mean dry yield kg/ha is 2127. Piperine % is 4.9, Oleoresin% 8.3 and essential oil % is 1.3. It is suitable for all the pepper growing tracks of India.

Panniyur -7 Open pollinated progeny selection from Kuthiravally. The mean dry yield kg/ha is 1410. Piperine % is 5.6, Oleoresin% 10.6 and essential oil % is 1.5. It is suitable for all the pepper growing tracks of India.

Subhakara	Selection from Karimunda (KS-27). Released from IISR, Calicut. The mean dry yield kg/ha is 2352. Piperine % is 3.4, Oleoresin% 12.4 and essential oil % is 6.0. It is suitable for all the pepper growing tracks of India.
Sreekara	Selection from Karimunda (KS-14). Released from IISR, Calicut. The mean dry yield kg/ha is 2677. Piperine % is 5.3, Oleoresin% 13.0 and essential oil % is 7.0. It is suitable for all the pepper growing tracks of India.
Panchami	It is a Selection from Aimpiriyan. It is a high yield clone released from IISR, Calicut.
Pournami	It is a Selection from Ottaplackal (Coll. 812)2333. It is tolerant to root knot nematode.
IISR Shakthi	Open pollinated progeny of Perambramundi. Tolerant to Phytophthora foot rot.
IISR Thevam	Clonal selection of Thevamundi. Tolerant to Phytophthora foot rot;Suited to high altitudes and plains.

Plant protection

Pests

Pollu beetle (*Longitarsus nigripennis*). It is the most destructive pest of black pepper and is more serious in plains and at altitudes below 300 m. The adult beetles feed and damage tender leaves and spikes while, females lay eggs on tender spikes and berries. The grubs bore into and feed on the internal tissues and the infested spikes turn black and decay. The infested berries also turn black and crumble when pressed. The term *pollu* denotes the "hollow nature" of the infested berries. The pest infestation is more serious in shaded areas and during September-October in the field.

Control: Effectively controlled by Endosulphan spray @0.05% twice (June and September)

Top shoot borer (*Cydia hemidoxa*): It is a serious pest in younger plantations. The larvae bore into tender terminal shoots and feed on internal tissues resulting in blackening and decaying of affected shoots. The pest infestation is higher during July to October when numerous succulent shoots are available in the vines.

Control

1. It is managed by Spraying quinalphos (0.05%) on tender terminal shoots. Repeat spraying at monthly intervals (during July-October).
2. Biological agents against this pest are *Hexamerimis sp., Apanteles cypris.*

Leaf gall thrips (*Liothrios karnyi*) The feeding activity of thrips on leaves causes the leaf margins to curl downwards and inwards resulting in the formation of marginal leaf galls. Later the infested leaves become crinkled and malformed. In severe cases of infestation, the growth of younger vines and cuttings in the nursery is affected.

Control: Spraying dimethoate (0 .05%) during emergence of new flushes in young vines in the field and cuttings in the nursery.

Biologically it can be controlled by *Montandoniola moraguesi* and *Androthrips flavipes.*

Diseases

Quick wilt or Foot rot or Root rot –It is caused by *Phytophthora capsici* pathogen infect leaves spikes, collar region and roots, collar infection cause sudden collapse of the vine. It is the most destructive of all diseases and occurs mainly during the south west monsoon season. All parts of the vine are vulnerable to this disease and the expression of symptoms depend upon the site or plant part infected and the extent of damage. If the main stem at the ground level or the collar is damaged, the entire vine wilts followed by shedding of leaves and spikes with or without black spots. The branches break up at nodes and the entire vine collapses within a month. If the damage is confined to the feeder roots, the expression of symptoms is delayed till the cessation of rain and the vine starts showing declining symptoms such as yellowing, wilting, defoliation and drying up of a part of the vine. This may occur during October-November onwards. These vines may recover after the rains and survive for more than two seasons till the root infection culminates in collar rot and death of the vine.

Control

1. Removal of the infected dead vines and burn them as this reduces the build up of inoculum (fungal population).
2. Provide proper drainage to avoid water stagnation
3. The freshly emerging runner shoots should not be allowed to trail on the ground. They must either be tied back to the standard or pruned off.

4. The nursery should preferably raised in fumigated or solarized soil.
5. After the receipt of a few monsoon showers (May-June), all the vines are to be drenched at a radius of 45-50 cm with copper oxychloride 0.2%. Spray vine with 1% Boredeaux mixture and 0.25% Ridomilziram.
6. Biologically it can be controlled by VAM fungi.

Slow decline or slow wilt –It is a debilitating disease of black pepper. Caused by soilborne fungi *Fusarium sp., Rhizoctonia sp., Pythium sp.* or Nematodes – *Radopholus similes, Meloidogyne incognita.* Foliar yellowing, defoliation and die-back are the aerial symptoms of this disease. The affected vines exhibit varying degrees of root degeneration due to infestation by plant parasitic nematodes. The diseased vines exhibit foliar yellowing from October onwards coinciding with depletion of soil moisture. With the onset of south west monsoon during May-June, some of the affected vines recover and fresh foliage develops. However, the symptoms reappear in subsequent seasons after the cessation of the monsoon and the diseased vines gradually loose their vigour and productivity.

Control

1. Severely affected vines which are beyond recovery should be removed from the plantation and destroyed.
2. Treat the planting pit with Carbofuran @ 50g at the time of planting.
3. Apply Carbofuran @ 100 g/vine twice in a year.
4. In areas severely infested with root knot nematodes, cuttings of the resistant variety 'Pournami' may be planted.

Pollu disease (Anthracnose) – It is caused by *Coletrotrichum gloesphorides.* It can be distinguished from the pollu (hollow berry) caused by the beetle by the presence of characteristic cracks on the infected berries. The disease appears towards the end of the monsoon. The affected berries show brown sunken patches during early stages and their further development is affected. In later stages, the discolouration gradually increases and the berries show the characteristic cross splitting. Finally, the berries turn black and dry.

Control: It can be controlled by the Spraying of 1% Boredeaux mixture.

Spike shedding

It is one of the emerging diseases especially recorded in variety Panniyur-1 at higher elevations. It is seen in serious condition when the pre-monsoon showers are delayed and flowering and spiking occur during June-Jul y. These spikes predominantly produce female flowers instead of bisexual flowers. Heavy spike shedding may occur due to lack of pollination.

Control: Irrigation of vines from second fortnight of March coupled with prophylactic spraying with Bordeaux mixture 1% or carbendazim 0.2% reduces the intensity of spike shedding.

Phyllody disease

This disease is caused by phytoplasma and the affected vines exhibit varying stages of malformation of spikes. Some of the floral buds get retransformed into narrow leaf like structures. Such malformed spikes show leafy structures instead of floral buds, exhibiting phyllody symptoms. In advanced stages, the leaves become small and chlorotic, and the internodes are also shortened. Severely affected vines become unproductive. In severely affected vines the entire spike is converted into small branches which appear chlorotic and the vines decline rapidly. The infected vine becomes unproductive within 2 to 3 years.

Control: Destruction of infected vines to prevent the further spread of the disease.

2.9 SMALL CARDAMOM

Botanical Name	*Elettaria cardamom*
Family	Zingiberaceae
Origin	Evergreen forest of Western Ghats in South India
Chromosome number	2n=48/52
Commercial value	Dried capsules especially the seeds
Chemical constituents	1, 8- cineole
Research Institute	Indian Cardamom Research Institute at Myladumpara.

Importance

Small Cardamom is one of the world's most ancient spices, was already mentioned in approximately 3000 BC in Sanskrit texts in India (Ravindran and Madhusoodanan 2002). It is popularly known as "queen of spices" and a precious ingredient in food preparations, perfumery, mouth freshners, health foods, medicines, beverages *etc.* as well as good for the digestive and respiratory system. India and Saudi Arabia consume more than half of the world's total cardamom production. In Arab countries and India, Cardamom is a common flavouring ingredient for coffee and tea. In Scandinavia, as well as in Germany and Russia, it is used to flavour cakes, pastries, and sausages. It is popular in Indian and South Asian cooking and used to make spice blends, such as curries and *garam masala*.

It is used for flavourings various preparation of food, confectionary, beverages and liquors. It activates the spleen and useful for hyperacidity and indigestion. It prevents nausea, and morning sickness. When eaten with banana, it controls vomiting. It is good for cough, cold, asthma and bronchitis. Chewing cardamom after a meal is recommended to aid digestion and to clean teeth. In Eastern medicinal practices it is used for curing such ailments as influenza, infections, asthma, bronchitis, cardiac disorders, diarrhea, nausea, cataracts, and for strengthening the nervous system. It is also said to have a cooling effect in hot climates. Ancient Greeks and Romans already used its delicate aroma to make perfume. Seeds of cardamom contains 6-8% essential oil but seeds of large cardamom contain less than 3.5% essential oil. Indian cardamom is slightly smaller, but more aromatic.

Distribution

Small cardamom is indigenous to South India, where it grows in the evergreen rain forests at an altitude between 700 and 1500 m. The major growing states are Kerala, Karnataka, Tamil Nadu. Around 80-85% of India's cardamom production is from Kerala-the high ranges of Idukki and Wayanad districts. Cardamom has been commercially cultivated in the Western Ghats for 150 years, owing to its sensitiveness to wind, drought, and water-logging, optimum yield is obtained on warm (10 to 35°C) and humid (with >1500 mm of well-distributed rainfall) mountain slopes at 600-1500 m elevation, under a canopy of evergreen trees. India had monopoly in area, production and export until 1979-1980. Later, Guatemala emerged as the premier producer and exporter accounting to about 90% of global trade. It is also grown in reasonable quantities in Sri Lanka, Papua New Guinea and Tanzania. At present, the largest producers of true cardamom are Guatemala and India, and smaller producers include Tanzania, Sri Lanka, Papua New Guinea, El Salvador, Laos, and Vietnam.

It is cheaper substitutes to real cardamom (*Amomum* spp. and *Aframomum* spp.) are grown and used in some Asian countries (Nair and Kutty 2004).

Description of Plant

It is a herbaceous perennial 2.4 m tall having branched subterranean underground rhizome which several leafy shoots arise, forming a clump. Its inflorescence is racemose. Flowers are bi-sexual and fragrant, fruits are trilocular capsule. Flowers initiate in March-April and reaches in full bloom in 30 days after flower initiation and bloom to maturity takes in 5-6months. Honeybees are the principal pollinating agent.

Cultivation

Climate

The natural habitat of cardamom is the evergreen forests of the Western Ghats. It requires early monsoon showers for flowering and fruit setting, They grow well in places receiving a well distributed rainfall throughout the year and with temperature ranging between 10° and 35°. It requires an annual rainfall of 1500-4000mm. It grows successfully upto 600-1500m above mean sea level. Summer showers are also required for panicle initiation otherwise it will affect the yield. The cardamom plant requires a continuous spell of rain interspersed with periods of good sunshine.

Soil

It is generally grown in forest loamy soils which are usually acidic (pH 4.5-5.8). Sandy clay to sandy loam texture is most suitable. Most of the cardamom growing regions are situated in heavy rain fall areas and therefore suffer from leaching of nutrients. This soil tends to be acidic due to leaching of bases. These soils are rich in organic matter and also in N, Fe and aluminum oxide but deficient in zinc, boron, molybdenum and manganese. They prefer well drained soils rich in organic matter. The crop does not tolerate water logging.

Cultivation

Propagation

It can be propagated by two ways seeds and Vegetative (through suckers each consisting of atleast one old and a young aerial shoot).

1. **Seed Propagation:** Seedlings are normally raised in primary and secondary nurseries. The nursery site should be selected on gentle sloppy lands, having an easy access to a water source. Raised beds are prepared after digging the lands to a depth of 30- 45 cm. The beds of 1 m width and

convenient length raised to a height of about 30 cm are prepared. A fine layer of humus rich forest soil is spread over the beds. Seeds should be collected from well ripe capsules. Immediately after the harvesting, the husk is removed and the seeds are washed repeatedly in water for removing the mucilaginous coating. After draining the water, the seeds are to be mixed with wood ash and dried in shade for a day. In order to ensure uniform and early germination, seeds should be sown immediately after extraction. If the sowing is delayed, pre-sowing treatment of seeds with 25% Nitric acids for 10 minutes is advisable to get a quick and higher germination. One kg of seed capsule may produce 5000 seedlings (600g seeds). Seeds are sown in November-January and germination is completed in 1 or 2 months. After germination, mulch is to be removed. Excess seedling is thinned out after 75-80 days of sowing. Thinned out seedling used for gap filling within the nursery bed or for raising secondary nursery. When seedlings attain 5-6 leaf stage, light earthing is to be done. In Kerala, Tamil Nadu the seedlings at 4-6 leaf stage are transplanted to the secondary nursery. Seedlings are transplanted in March-May at a spacing of 20 × 20 cm.

2. **Vegetative Propagation:** Vegetative method is becoming popular as this ensures large scale production of true to type planting material of high yielding lines. The suckers free from pest and disease are better suited for clone multiplication of high yielding varieties. Suckers, each consisting of one grown up tiller and a growing young shoot, are placed at a distance of 0.6 m in the trenches during March-October. Regular cultural operations are to be followed including high fertilizer dosed 100:50:200 kg NPK/ha in 6 split doses at 60 days interval along with Neem cake at 250 g/plant. Irrigation should be provided atleast twice in a week. Overhead pandal at a height of 3.6 m covered with coir mat or leafy twigs of any shade tree may be provided during the non-rainy season. Within a period of 12 months, a plant would produce atleast 32 to 42 suckers, which may yield atleast 16 to 21 planting units per ha of clonal nursery within 12 months of planting.

Method and Time of Planting

10-18 months old seedlings are used for planting in the field. Dig the pits at 45 × 45 × 30 cm in April-May and fill with top soil, compost or FYM. Planting 2×1m is advisable along the contour. For Malabar variety 2m × 2m spacing is advisable. Ideal planting time is in the month of June is suitable due to cloudy days with light drizzle.

Fertilizer and manures

Recommended dose of fertilizer for small cardamom is 75: 75: 150 kg NPK per ha along with Neem oil cake @ 1 kg/clump and Cattle manure @ 5 kg/clump. In trench system with high density planting the application of 120, 120 and 240kg of NPK/ha/year resulted high productivity. Time of application of fertilizers is May and September.

Under rainfed condition, fertilizer should be applied in two splits. First application helps in production of suckers and develops of capsules while second helps in initiation of panicles and suckers. In irrigated condition fertilizers in four split doses of quarterly interval is beneficial. Cardamom being a pseophyte, is very sensitive to moisture stress. Shade helps to regulate soil moisture as well as temperature and provide congenial microclimate to cardamom. .

Irrigation

In order to overcome the dry spell during summer, it is necessary to irrigate the crop to get maximum production as it helps in initiation of panicles, flowering and fruit set. It is quite essential to irrigate cardamom plants from last week of January to mid May.

Mulching

It is an important cultural practice in cardamom. Fallen leaves of the shade trees are utilized for mulching. Sufficient mulch should be applied during November - December to reduce the ill effects of drought, which prevails for nearly 4 to 5 months during summer. Exposing the panicle above the mulch is beneficial for the pollination by bees.

Trashing

Trashing consists of removing old and drying shoots of the plant once in a year with the onset of monsoon under rainfed conditions and 2-3 times in high density plantations provided with irrigation facilities.

Harvesting

Cardamom plants normally start bearing two years after planting. Fruits mature at 30-40 days intervals. Throughout the cropping season of cardamom, *i.e.* from August to March approximately 6 picking is done in each 45 days interval. In most of the areas the peak period of harvest is during October-November. Ripe capsules are harvested in order to get maximum green colour during curing. The output of cardamom is greatly influenced by climatic conditions. Capsule is harvested just short of full ripeness. Average yield of dry capsules is 500 kg./ ha. After harvesting, capsules are dried. Green cardamom capsule soaked in 2% washing soda for 10 min. prior to drying helps to retain the green colour.

Processing

After harvest, capsules are dried either in fuel kiln or in electrical drier or in the sun. It has been found that soaking the freshly harvested green cardamom capsules in 2% washing soda for 10 minutes prior to drying helps to retain the green colour during drying. When drier is used, it should be dried at 45 to 50°C for 14 to 18 hours, while for kiln, over night drying at 50 to 60°C is required. The capsule kept for drying are spread thinly and stirred frequently to ensure uniform drying. The dried capsules are rubbed with hands or coir mat or wire mesh and winnowed to remove any foreign matter. They are then sorted out according to size and colour, and stored in black polythene bag to retain the green colour during storage. These bags are then kept in wooden chambers.

Plant protection

Pest

Thrips (*Sciothrips cardamomi*)– Adult and Nymphs suck sap and cause damage to leaves, shoots, inflorescence, etc. Affected capsules fetch lower price.

Control

1. Regulate shade in thickly shaded area.
2. Spray Monocrotophos 0.025% during March to September.

Shoot panicle and capsule borer (*Conogethes punctiferalis*)- Larvae bore the unopened leaf buds, panicles causing drying of feed on young seeds causing the capsules empty.

Control

1. Spray monocrotophos @0.075% at early stage of infection.

Aphids (*Pentalonia nigronervosa*)- Nymphs and adults suck the sap and act as vector of the mosaic or 'Katte' Virus.

Control

1. Spray 0.05% Dimethoate.

Nematode (*Meloiodogyne incognita*) - It occurs in nursery and main field. Poor germination and establishment in the nurseries. Stunted and poor growth of plants having galling and abnormal branching of roots.

Control

1. Treat the plants in the nursery with carbofuran 3 g @ 5 kg i.e. /ha or in the main field with carbofuran 5-g a.i./clump and apply 0.5kg of Neem cake per clump twice a year.

Diseases

Katte diseases (Viral disease) - Spindle shaped, slender chlorotic flecks appear on youngest leaves, later there develop into pale green discontinuous stripes as leaves mature, mosaic symptoms are marked, infected clumps are stunted, smaller in size, with slender tillers and shorter panicles.

Control: Use healthy seedlings, rogue the infected plants.

Azhukal or capsule rot – It is caused by *Phytophthora meadii* (*P. nicotianae* var. *nocotianae*). Occurs in heavy rain zone area. It affected capsules which turn brownish black in colour.

Control

1. It can be effectively controlled by removing infected and dead plants.
2. During pre-monsoon months, spray 1% Bordeaux mixture during May and August.

Damping off or rhizome rot- It is caused by *Pythium vexans* and *Rhizoctonia solani*. Excessive soil moisture and poor drainage favours the disease in nursery, infected seedlings collapse at collar region and die in patches, entire clump dies in grown up plants.

Control

1. Pre treatment in the nursery with 1:50 formaldehyde, drench the soil after germination with 0.2% copper oxychloride.

Varieties

Based on the nature of panicle and shape and size of fruits following types have been identified.

Malabar Type	It adapts well in lower elevation. It is relatively tolerant to water stress or drought. Plants are pale golden yellow
Mysore Type	It adapts well in higher elevation. It prefers well distributed rain. Plant stature is tall (3-5m). Capsule colour is green.
Vazhukka Type (natural breed between Mysore and Malabar)	It adapts well in higher elevation. Prefers well distributed rain. Plant stature is tall (3-5m). Capsule colour is green

New improved cultivars of cardamom

Coorg Cardamom Selection-1	Malabar type which released by IISR, Cardamom Research Centre, Appangala. It is an open seedling of clone and produces 408kg dry capsules / ha. Dry recovery 22%, Essential oil content: 8.7%.
Mudigree-1	Malabar type, Released by RRS, Mudigree. It is suitable for Malanad region of Karnataka state. It produces 250 to 300kg dry capsules / ha. Dry recovery: 20%, Essential oil: 8%.
IISR, Kodagu-Suvasini	It is Malabar type which produces 745 kg/ha dry capsules and is produced suitable for Karnataka, dry recovery 22%, essential oil 8.7%.
IISR-Avinash	It is Malabar type with dry capsule yield 847 kg/ha. Suitable for Karnataka and Wayanad, dry recovery 20.8%, essential oil 6.7%.
ISR-Vijetha-1	It is Malabar type and produces 643 kg/ha dry capsules and is suitable for Karnataka and Wayanad, dry recovery 20.8%, essential oil 7.9%.

References

Ravindran P.N. and Madhusoodanan K.J., editors. 2002. Cardamom: The Genus Elettaria. Medicinal and Aromatic Plants—Industrial Profiles. London, United Kingdom: Taylor and Francis.

Nair T.K.R. and Kutty M.G. 2004. Cardamom (*Elettaria cardamomum*) in Kerala, India. In: Kusters K, Belcher B, editors. Forest Products, Livelihoods and Conservation: Case Studies of Non-timber Forest Product Systems. Vol 1: Asia. Bogor, Indonesia: CIFOR [Center for International Forestry Research], pp 133–150.

2.10 LARGE CARDAMOM OR NEPAL CARDAMOM OR GREATER INDIAN CARDAMOM

Botanical Name	*Amomum subulatum* Roxburg
Family	Zingiberaceae
Origin	Eastern Himalaya Region
Chromosome number	2n=48
Commercial value	Pods
Chemical constituents	α- Terpinyl acetate
Research Institute	Indian Cardamom Research Institute (ICRI) at Myladumpara (Kerala)

Importance

Large cardamom is used as a spice and also in several Ayurvedic preparations. It contains 2–3% essential oils, possesses carminative, stomachic, diuretic and cardiac stimulant properties and is also a remedy for throat and respiratory trouble. Large cardamom has a pleasant aromatic odour, due to which it is extensively used for flavouring vegetables and many food preparations in India. It is also used as an essential ingredient in mixed spices preparation having medicinal value. The decoction of seeds is used as a gargle in infection of teeth and gums. Large cardamom seeds are considered as an antidote to either snake venom or scorpion venom.

Distribution

Large cardamom commonly known as '*badi elaichi*' is one of the world's very ancient spices. It grows in the wild and is also domesticated in the sub-Himalayan region, at altitudes ranging from 1000 to 2000 m above mean sea level. It is one of the main cash crops cultivated in Sikkim, Nagaland, Uttaranchal, Darjeeling and some other parts of the North east India.

Description of plants

The plant is a perennial bush. It has a sheathed stem, reaching 2–5 m in height with a large tuberous rhizome and leaves that are 30-60 cm long and 5–15 cm wide. The trailing leafy stalk that grows from the plant base at ground level bears the seed pod. The flowers are green with a white-purple vein tip. The small, brown-black sticky seeds are contained in a pod in three double rows with 25–70 seeds in each row, depending on the cultivar. The pods are roughly triangular in cross section and oval or oblate. The dried surface is rough and furrowed with deep wrinkles. Pods are available whole or split, and the seeds are sold loose or ground.

Cultivation

Climate

Large cardamom is a shade loving plant (pseophyte). Its natural habitat is humid subtropical, semi evergreen forests on steep hills of eastern sub-Himalayan region. It can be successfully grown at 6000-2350 mean sea level or above. Normally cultivated in lower altitudes of cooler areas and high attitude of warmer areas. During severe winter its plants remain dormant and withstand up to 2°C, but frost hailstorms are injurious. The flower bud differentiation occurs from August in lower altitudes to October in higher altitudes. The initiation of flower buds occur before winter on lower elevations but further development takes place only after the lapse of cold period in early-March. The continuous rain during flowering is detrimental as it hampers foraging activity of the pollinating bees affecting the sensitive flowers and resulting in poor capsule setting and barren spikes.

Soil

It requires moist but well-drained, loose sandy loam to loamy soils. Its crop requires at least moderate deep (6m) soil for its good performance. The soils should have a pH range of 4.5 to 6.0. Even though the crop can be grown in undulating and steep terrains, land with a more moderate slope is preferred. Due to high rainfall conditions, the soils are mostly acidic and have high exchangeable iron and aluminum.

Propagation

Propagation of large cardamom is done through seeds and suckers but commercially it is propagated through suckers. Propagation through seeds enables the production of a large number of seedlings. Virus diseases are not transmitted through seeds and therefore the seedlings are free from viral diseases, if adequate care is taken to isolate and protect the nursery from fresh infection. Plants raised from seeds need not necessarily be high yielders even if they are collected from very productive plants due to cross-pollination. The major pollination is done by wild bees and the rest by honey bees. Planting suckers on the other hand ensures true to type and high productivity if they are collected from high yielding plants (Pathak, 2008).

Primary nursery

Cardamom seeds are generally sown in September-October. Seed beds are prepared in well drained soil dug to a depth of 30 cm and left for weathering. Raised beds with 15 to 25 cm height, 1 metre width and convenient length, preferably 6 metres, are prepared. Well decomposed farm yard manure is mixed with the soil and the surface of the bed is made to a fine tilth. 80-100g of seeds are sown per bed in lines spaced 10 cm apart. The seeds are then covered with fine soil and mulched with paddy straw/dry grass (10-15 cm thick). The beds are watered regularly and weeding is done as and when required. When the seedlings attain 3-4 leaf stage, they are transplanted to secondary beds.

Secondary nursery

Beds of size 15 cm in height and 100 cm in width with convenient length are prepared and well-decomposed farmyard manure is mixed with the soil and an even surface is formed. Seedlings with 3-4 leaves are transplanted to the beds in May-June at spacing of 15 cm between them. An overhead *pandal* is erected for providing shade and the soil is kept moist with irrigation. When the seedlings attain a growth of 45-60 cm in height with 2-3 tillers, they are planted in the main field during June-July of the subsequent year (Pathak, 2008).

Sucker multiplication nursery

For commercial propagation suckers are used. They should be generated only in sucker multiplication nursery where adequate precautions are taken to ensure that viral diseases are not transmitted through the suckers produced. The site should be located at least 500 metres away from large cardamom plantations and established either under the shade of forest trees or under shade *pandals* with 50% shade.

Trenches of size 30×30 cm are prepared at convenient lengths with an inter space of 30 cm. Well decomposed farm yard manure or compost is mixed with the soil and filled in trenches. Then the suckers from high yielding disease free plantations, with one grown up shoot with an emerging bud are planted at 30 cm in the trenches. The time for planting is May-June. After planting, the plant base is mulched with dried forest leaves. The multiplication rate in this method is about 1:8 in one years time. The grown up tillers are split into units of one tiller with an emerging bud and planted in the main field during June-July.

Planting

The plantation site should be preferably near a perennial water source. Keep only the appropriate shade trees. A layout for foot-paths, irrigation channels or for drainage is to be carried out. It is necessary to establish shade prior to planting of cardamom. Hence planning and planting of shade trees saplings should be made well in advanced by at least 2 years. Keep only appropriate shade by thinning out excess branches so that only 35% sunlight is in filtered down to its canopy. The shade trees are planted either in alternate rows or mixed intermittently in the same row with a spacing of 7-10 m. between plants for uniform shade on plantations. Planting is done during June-July when there is enough moisture in the soil. Pits of size 30×30×30 cm are prepared on the contour of the hill at a spacing of 1.5×1.5 m after the onset of monsoon showers. A wider spacing of 1.8×1.8 m is recommended for robust cultivars like Sawney, Varlangey, Ramsey, etc. The pits are left for weathering for a fortnight and then filled with topsoil mixed with cowdung or compost at the rate of 1-3 kg. per pit. Seedlings/suckers are planted in the middle of the pit. Care should be taken not to plant the seedlings/rhizomes very deep in the pit. After planting the seedling/suckers may be staked and the base of the plant is mulched with dry leaves (Pathak,2008).

Fertilizer and Manure

Provide forest soil or leaf litter @ 5 kg mixed with low graded composites NPK fertilizer (15: 10: 12) @ 2kg/plant during April-May.

Irrigation

Cardamom does not tolerate drought. A continuous dry spell of about 2 months may result in huge mortality. Constant maintenance of optimum soil moisture ensures better growth and early bearing. Irrigations may be provided once in every ten days during, December-April.

Shade-regulation

The crop grows well in shaded areas. Tall growing trees are beheaded regularly at a height of 4-5m. to encourage spreading habit with renewed vegetative vigour to provide a uniform shade and higher quantity of leaf litter.

Roguing and gap filling – About 5-10% of plants are damaged due to hailstorm, disease insect pests etc. Hence remove them and replace them by healthy one. Ideal time for gap filling is May-July.

Harvesting

Harvest the crop at right maturity. It is harvested before it ripens to avoid the capsules from splitting during the drying process. Do not delay removing capsules from the spikes after harvest. Remove capsules from the outer and middle whirls and collect them in one lot and those from the inner whirls in another lot. The best time of harvesting at low altitudes is during August-September whereas, at high altitude it is November-December. Sustained yields are normally expected up to fifteenth year from planting. However, constant replanting by replacing the old and degenerated clumps every year in the plantation along with optimum shade and nutrition may keep them productive for many more years.

Post- Harvest

Raw capsules contain up to 85% moisture and hence are of poor keeping quality. Therefore they are immediately cured to retain only about 10-13% moisture on dry-weight basis. The retention of natural colour and flavour during curing is most important. The flavour constituents are highly volatile and are easily lost due to direct heat and higher temperatures (>55°C). Therefore it is necessary to adopt appropriate curing techniques. Remove extraneous matter and wash thoroughly the harvested produce before drying. Adopt flue pipe system of curing to retain the original colour, aroma and flavour. Dry the capsules immediately after harvest. Dried cardamom should not contain more than 10% moisture for better shelf life. Remove tails after drying.

Curing of Raw Cardamom

Curing process involves indirect heating of the raw capsules at an optimum temperature range of 50-55°C which is followed by rapid air circulation within and afterwards exhaustion of moist air from the curing unit.

Traditionally large cardamom is cured in a *bhatti* where the capsules are dried by direct heating. Under this system the cardamom comes in direct contact with smoke which turns the capsules to a darker browner black colour with a smoky smell. Improved curing techniques are available by which cardamom is processed to give better quality and appearance. One such method is the ICRI Spices Board improved ***bhatti*** system of curing in which the cardamom is dried by indirect heating at 45-50°C.

Packing of Cured Cardamom

Grade according to size and colour and store in moisture are proof containers. Use polythene lined gunny bags for storing. Store in dry places and in wooden boxes. Take necessary precaution against rat and insect damages. The dried cardamom may be preferably sold soon after drying provided the growers get remunerative price, as long storage may lead to deterioration of quality. Packing is done in insect-proof bags. Coal-tar coated and polythene lined gunny bags are effective against insects during storage. Curing of capsules up to a moisture level of 10% and packing them in air-tight containers also ensure protection against insects.

Yield

About 450-500 kg dry or cured capsules are obtained from 1 ha of land during 4th-10th year.

Plant Protection

Pests

Large cardamom is large free from the attack of any major pests except for the sporadic incidence of leaf eating caterpillars. Aphids are found in most of the areas which transmit the viral diseases chirke and foorkey.

Disease

Large cardamom is susceptible to several viral and fungal diseases. But fungal diseases are not the major problem. The major disease of widespread occurrence is viral diseases *viz*; Chirke and Foorkey.

Chirke disease

It causes mosaic like symptoms in leaves.Pale streaks are observed in leaves. The streaks turn brown and causes leaves and plant to dry and wither. The flowering in diseased plants is extensively reduced and only 1-5 flowers develop in the inflorescence as against 16-20 of a healthy plant. Growth and yield of the affected plants gradually declines and ultimately they perish. By the end of third year after planting the loss can be as much as 85%. The disease is transmitted by aphids (*Rhopalosiphum maidis* Fitch.) within a short acquisition period of 5 min. It also spreads by planting infected suckers. Transporting of infected suckers from one area to another leads to the spread of this disease. The disease is also transmitted mechanically through the knife used for harvesting.

Foorkey disease

It is characterized by dwarf tillers with small, slightly curled pale-green leaves. The virus induces remarkable reduction in the size of leafy shoots and of leaves of the infested plants and proliferate large number of stunted shoots arising from the base (rhizome). The inflorescence is transformed into leafy vegetative parts along with suppression of fruit formation altogether. The diseased plant remains unproductive and degenerates gradually. The primary spread of the disease is through infested rhizomes and further spread within the plantation is through aphids *viz.,Pentalonia nigronervosa* Cog. and *Micromyzus kalimpongensis* Basu.

Management of Chirke and Foorkey diseases

Plants affected by the viral diseases cannot be cured but the losses can be minimized by adopting appropriate management practices.

A) Treatment of Infected plants

i. Keep a constant vigil to detect disease affected parts.

ii. Uproot and destroy affected plants as soon as symptoms appear. Repeat detection and uprooting at regular intervals.

iii. Uprooted plants should be taken to isolated places chopped into small pieces and buried in deep pits for quick decomposition.

iv. Infected rhizomes could be killed by injecting Agroxone-40.

B) Prophylactic measures

i. Use seedlings produced in certified nurseries.

ii. Propagation through suckers is recommended only through certified multiplication nurseries.

iii. Avoid raising nurseries in the vicinity of infested plantations.

iv. Avoid replanting suckers from areas that are prevalent in disease.

Varieties

1. Bebo : Drought tolerant variety of large cardamom; resistant to foorkey viral infection; capsules are medium to bold in size; seeds contain 2% essential oil.

2. Varlangey : It is grown in mid and high altitude (> 1500 m above mean sea level) areas. Its yield performance is exceptionally high at high altitudes. It is a tall variety and plants grow up to 2.8m; yields big, bold and elliptic capsules; 15–20 capsules per each spike and 40–60 seeds per each capsule. Flowering starts in March and harvesting is done in September-October. It yields 500kg of cured capsules/ha. Essential oil in seeds is 2.5%.

3. Golsey : Self-shaded, less vigorous growing variety of large cardamom; grown at low and mid elevations; clump size is smaller with less number of tillers and hence suited to close spacing; capsules are bold, roundish with reddish-brown to dark-pinkish and slightly echinated; consistent performer though not a heavy yielder; immune to chirkey but not to foorkey disease; susceptible to fungal leaf-spot disease; on an average, capsules contain about 40–55 seeds; volatile oil content of the seeds is 2–3.0%.

4. Ramla : Tall (1.5-2.0 m) and vigorous; capsules are dark pinkish, uniform and medium-bold in size having 25–35 seeds each; seeds contain 2.2–3% volatile oil content; moderately tolerant to chirkey but susceptible to foorkey disease. Flowering commences in May and the crop is generally ready for harvest in October.

5. Ramsey : It is well suited to high altitudes above 1500 meters above mean sea level and can be cultivated even in steep slopes. Tall (1.5-2.0m), vigorous, slanting with drooping leaves. It occupies a major area under cultivation in Sikkim and Darjeeling. Capsule size is small with less number of seeds (16–30), the colour being reddish-brown to maroonish. The essential oil content in seeds is about 1.5–1.8% on volume by weight basis, susceptible to chirkey and foorkey especially at lower altitudes. Flowering starts in May and crop is ready for harvest by October-November.

6. Sawney	:	It is a widely adapted cultivar, which is most suited to medium (975 - 1500m above mean sea level) and high (> 1500 m above mean sea level) altitude areas. Vigorous-growing with maroonish to greenish-brown leafy stems slightly tapering towards apex. Capsules are reddish-brown to maroonish, variable in size from medium to bold with volatile oil content of 1.8–2.5%, plants are susceptible to both viral and leaf-spot and spike-rot diseases; heavy yielder.
7. Seremna	:	The cultivator is grown in a small pocket of the Hee-Goan, West Sikkim at low altitude and is known for its high yield potential. Plants are 1.5 to 2.0 meters tall, tillers are green and leaves are mostly drooping type, hence named as "Seremna" On an average 2 3 spike in each productive tiller with average 10 capsules in each spike and 65-70 seeds per capsules are recorded.
High Yielding released varieties	:	There are two high yielding varieties released by Indian Cardamom Research Institute in the year 2004 for cultivation in Sikkim and Darjeeling. The following are the varieties.
ICRI Sikkim- 1 ICRI Sikkim- 2	:	These varieties are selected from Sawney and now are mostly used in Tissue culture for planting material production.

References

Pathak, A. (2008) Cultivation of Large Cardamom in Sikkim. Facets of the North-east. pp:3

Varieties of Large Cardamom www. http://agrihortico.com/ and http://www.sikkimagrisnet.org/General/en/Cardamom_Varieties.aspx

2.11 CUMIN/ *ZEERA*

Botanical Name	*Cumimum cyminum*
Family	Apiaceae
Origin	Mediterranean region (Syria)
Chromosome number	2n=14
Commercial value	Seeds
Chemical constituents	Cuminaldehyde (4-isopropyibenzaldehyde)
Research Institute	National Research Centre on Seed Spices, Ajmer

Importance

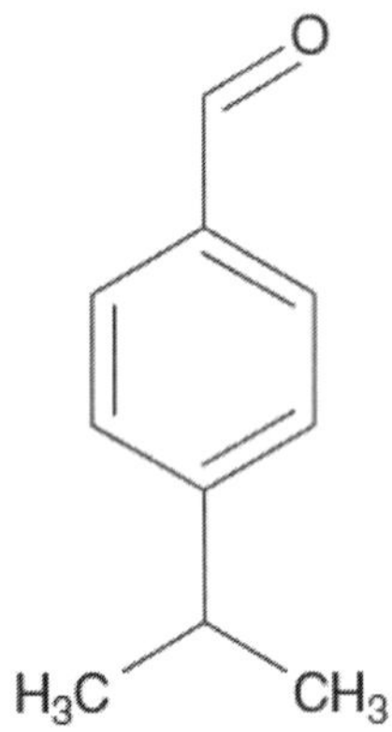

Cuminaldehyde

Cumin is known for its antioxidant properties. It is an important spice mainly cultivated for flavouring vegetables, pickle's and soups. The seeds are extensively used in Ayurvedic medicines prescribed for stomach pain and dyspepsia. It stimulates the appetite and a great herb for digestive disorders with some antiseptic properties. The seeds themselves are rich in iron and are thought to help stimulate the secretion of enzymes from the pancreas which can help absorb nutrients (vitamins and minerals) into the system. It has also been shown to boost the power of the liver's ability to detoxify the human body. Like most of the aromatic spices, Cumin is very rich in iron (above 66 mg. in each 100 grams) which is more than 5 times the daily requirement of iron for an adult; thus very good for pregnancy or lactating mothers, women who are undergoing menses, along with growing children and adolescents. It helps to strengthen nails and hair, making hair shiny and glossy. It even has remarkable amount of calcium (above 900 mg per 100 grams) which accounts to over 90% of our daily requirement of calcium. Recent studies have revealed that cumin seeds might also have anti-carcinogenic properties. Among the seed spices, cumin fruits have a distinctive bitter flavour and strong warm aroma due to their abundant essential oil content. Of this 40-65% is cuminaldehyde (4-isopropyibenzaldehyde), the major constituent and important aroma compound, as also the bitterness compound reported in cumin. The characteristic flavour of cumin is probably due to dihydrocuminaldehyde and monoterpenes.

Distribution

The spice is native to Syria where cumin thrives in its hot and arid lands. Cumin seeds have been found in some ancient Syrian archeological sites. The word cumin is derived from the Latin *'cuminum'* and Greek *'kuivov'*. The Greek term itself has been borrowed from Arabic. Forms of this word are attested in several ancient Semitic languages, including *'kamunu'* in Akkadian. The ultimate source is believed to be the Sumerian word *'gamun'* (Zohary and Hopf, 2000). It is grown mainly in hot countries, especially India, North Africa, China and America. In India it is exclusively cultivated in Rajasthan and Gujarat. Banaskantha and Mehsana in Gujarat, and Barmer, Jalore, Jodhpur in Rajasthan are the major cumin producing area. West Bengal, Uttar Pradesh, Andhra Pradesh and Punjab also make significant contribution to Indian output (Anonymous, 2009).

Description of Plant

It is small and slender, much branched, annual herb, about 25cm in height, leaves linear, bluish green, flowers white or pink in small compound umbels. Fruits are grayish. Plants come into bearing in 60-100days from flowering.

Cultivation

Climate

Cumin is tropical plant. It grows well in sub-tropical climate too. Moderately cool and dry climate is best for cumin cultivation. It is not suitable in the areas with high humidity and frequent rains, particularly after flowering period, as it increases the incidence of disease like blight and powdery mildew.

Soil

Cumin can be cultivated in all types of soils but well drained sandy loam and medium soils are suitable for the crop. It can also be cultivated on medium heavy as well as moderately saline soils with good drainage.

Land Preparation

Soil is brought to fine tilth by 2-3 ploughing followed by harrowing. Stubbles of previous crops should be collected and removed from the field. Clods should be broken and field should be leveled with the help of plank. Beds of 4× 3 m size with provision of irrigation channels should be prepared before sowing of seeds to facilitate proper irrigation and intercultural operations.

Sowing time and Seed rate

Sowing is done from 1st week of November to 1st week of December by broadcasting or in rows drilled at 30 cm. Seed rate vary from 12 to 15 kg / ha, depending upon method of sowing & type of soil. Sowing of seeds should be done at a depth of 1 – 2 cm after treatment with Ceresan or thiram or Difoltan @ 3.0 g per kg. Soaking of seeds for 8 hours before sowing is helpful in getting good germination. Soaked seeds should be dried in shade to facilitate broadcasting. Sowing at higher depths affect the germination of seeds, adversely. Crop rotation should be followed to avoid incidence of pest and diseases.

Manure and fertilizer

About 10-15 tonnes FYM, 30 kg Nitrogen and 15 kg Phosphorus per ha is recommended for this crop. Full dose of FYM should be mixed into the soil at the time of land preparation and 15 kg Nitrogen and whole quantity of Phosphorus should be applied as basal dose for proper growth and development. Another, 15 kg Nitrogen should be applied as topdressing one month after germination of seeds.

Weed control

The crop should be free from weeds because it faces severe crop-weed competition because of its slow growth and short stature. Hence twice weeding and hoeing is necessary at 30 and 60 days after sowing. In drilled crop light intercultural operation is beneficial. Application of fluchloralin @ 1 kg/ha is effective to control weeds.

Irrigation

Based on type of soil, crop requires 4-6 irrigation. Immediately after sowing 1st irrigation should be given followed by second irrigation at 6-10 days interval. Subsequent irrigations should be given after 30, 45, 65 and 80 days from 1st irrigation. Irrigation at the time of flowering and fruit set are essential. At maturity stage irrigation should be stopped.

Harvesting

Cumin crop takes about 110-115 days to reach maturity. It becomes ready to harvest, when plants turn yellowish brown. Harvesting should be done early in the morning by cutting/uprooting the whole plants. Plants are picked up for drying in the sun and after drying seeds are separated by light beating with sticks. Seeds should be cleaned by winnowing.

Yield

8-10 quintals per ha

Plant Protection

Diseases

Fusarium wilt: This fungal disease in severe in young plants. Infected plants show peculiar symptoms of dropping of tips and leaves, leading to mortality of the entire plant.

Control

1. Crop rotation and use of Neem cake are helpful in checking spread of the fungus.
2. Seeds collected from disease free plots should only be used for sowing.

Alternaria Blight: The blight affected plants show very minute brownish necrotic spots, which later turn to blackish. Mostly diseased plants fail to produce seeds. If seeds are produced they remain shriveled, light in weight and dark in colour.

Control

1. Seed treatment and spraying of 0.2% solution of Dithane-M-45, four times at 10 days interval commencing from 40 days after sowing is recommended.
2. The crop should be kept free from weeds.
3. Mustard should not be grown in the vicinity of this crop.

Powdery mildew: This disease affected plants in early stages. Diseased plant show minute whitish spots on leaves, petiole, stem pedicel and seeds. In severe condition, it looks as the plants have been dusted with white powder. At later stages of attack seeds become white and shriveled and light in weight.

Control

1. Crop should be dusted with 300 mesh Sulphur dust @ 25 kg/ha to control this disease as soon as the symptoms are noticed.
2. Spraying of wettable Sulphur or Dinocap (Kerathan or Thiowet) can also be used to control the disease @ 20-25 g per 10 liter of water at the initial stage of this disease. If needed second spray should be given 15-20 days after first spray.

Pest

Aphid: This is a major pest of cumin crop, it sucks the sap of tender parts and reduces the yield.

Control

1. Spraying of 0.03% solution of Dimethoate or 0.025% solution of Methyl demetone or 0.04% solution of Monocrotophos is recommended to control the aphid.

Leaf eating Caterpillar: This pest causes damage to the foliage of plants reducing yield of the crop.

Control

1. Spray 0.02% solution of Phosphamidon in the early stage of crop.

Varieties

Improved varieties like S-404, MC-43. Gujarat Cumin-1(GC-1), GC-2, GC-3, RS-1, UC-198, RZ-19, etc., evolved by Agricultural Universities of Gujarat and Rajasthan having higher yield potential are useful for cultivation. The maturity days varies from 110 to 115 days depending on variety.

References

Anonymous (2009). Short Term report on Cumin (*zeera*) October 2009. Indiabulls Commodities Limited.

Zohary, D. and M. Hopf. 2000. Domestication of plants in the Old World. 3rd Ed. Oxford University Press, p. 206.

2.12 NIGELLA/ BLACK CUMIN/ MANGARAIL/ KALONGI

Botanical Name	*Nigella sativa* L.
Family	Ranunculaceae
Origin	Levant in the Eastern Mediterranean region.
Chromosome number	2n=10
Commercial value	Dried seed and oil
Chemical constituents	Nigellin

Importance

It is a minor seed spice crop of India. It has been employed for thousands of years as spice and food preservative, as well as a protective and curative remedy for numerous disorders. It is a significant source of protein, essential fatty acids (the type your body can't produce), and many vitamins namely, A, B, B2, C and Niacin 21% Carbohydrates, 35% and plant fats 35-38%. Minerals include Calcium, Potassium, Iron, Zinc, Magnesium and Selenium. Seeds are used in the preparation of pickles, as one of the ingredients. The seeds contain a brown coloured volatile oil 0.5 % to 1.6 % and red coloured stable oil which is 31 %. Besides this it contains albumin, sugar, carbonic acid, seponin, melanthin, Arabic acid, a bitter compound named nigellin, resins, tannins and ash 7 %. It contains a volatile oil carvone 45 to 60 %, D-lymonine and cymine. It also contains nigellone, which is helpful in supporting the respiratory tract. It is known to have many medicinal properties in traditional medicine. It is used for different treatment such as Skin related problems, hair fall, headache, rhinitis, piles, pain, tastelessness, indigestion, abdominal distension, cough, asthma, pains in flanks, painful parturition, decrease milk secretion in lactating mothers, dysuria and fever.

Distribution

Nigella is grown in Egypt and Turkey. These regions are considered by many to be the source of the best nigella in the world. Wild growth of this plant is widespread in southern Europe, Northern Africa and Asia. In India, it is cultivated commercially in Punjab, Madhya Pradesh, Bihar, and Assam. Nigella producing countries are Pakistan, Sri Lanka, Bangladesh, Nepal, Iraq and Egypt.

Description of Plant

An herbaceous plant of annual growth, belonging to the buttercup family, about 30-60cm high. The greyish green leaves are wispy and thread like. The flowers are solitary having five petals about 2.5 cm wide, white with blue veins and the plant appears between June and September. Nigella seeds are small, black grains with a rough surface and an oily white interior. They are roughly triangulate (three sided), and approximately 1-3 mm long. The seeds have no natural smell although when they are rubbed they give off an aroma reminiscent of oregano. They have a strong, spicy / peppery taste, are slightly bitter, with a crunchy texture.

Cultivation

Climate

It is a cool season crop and is cultivated in the northern plains during winter season as a *Rabi* crop. Cold weather favours for early growth period and crop requires warm sunny weather during seed formation and maturity.

Soil

Loamy, medium to heavy soils with better fertility level are most suitable.

Seed rate and spacing

7 kg seeds are required per hectare. Spacing of 30 ×10-15cm is required.

Manure and Fertilizers

10-15 tonnes/ha FYM along with 40kg N and 30kg P_2O_5/ha should be applied for proper growth and development.

Harvesting

Nigella crop is ready for harvesting in 140-160 days after sowing.

Yield

Average yield is 0.8-1.0 tonnes/ha.

Plant protection

Disease

Root rot: It is a major problem in the cumin cultivation. It can be effectively controlled by Trichoderma (4g/kg seed).

Pest

Aphid: The infestation can be minimized by spray of 0.03% Dimethoate (30EC) or 0.1% Malathion (50EC).

Varieties

AN-1	Developed by NRC on seed spices, Ajmer, India. It possesses resistance to seed rot.
Azad Kalongi	The seed yield is 1.0-1.1tonnes/ha.

2.13 FENNEL / *SAUNF*

Botanical Name	*Foeniculum vulgare* Mill.
Family	Umbelliferae/ Apiaceae
Origin	Southern Europe and Mediterranean region
Chromosome number	2n= 22
Commercial value	Seeds
Chemical constituents	Terpenoid anethole, found in the volatile oil
Research Institute	National Research Centre on Seed Spices, Ajmer

Importance

The word "fennel" developed from the Middle English *fenel* or *fenyl*. This came from the Old English *fenol* or *finol*, which in turn came from the Latin *feniculum* or *foeniculum*, the diminutive of *fenum* or *faenum*, meaning "hay". Fennel oil (Essential oil) extracted from fennel seeds is used for manufacturing cordials and as a fragrant agent in toiletries such as soaps and shampoos. Fennel oil is extensively used as a flavoring agent in baking and confectionary industries. It is useful in the treatment of anemia. Fennel is one of the best remedies for curing indigestion and is used as an anti flatulent agent. It is extensively used for treating constipation, diarrhea, renal colic, respiratory disorders, and menstrual disorders. Fennel water is commonly given to infants as medicine. The root of fennel plants may be used as a purgative. Fennel seeds are used as stimulant and as a carminative. Fennel is extensively used in cure of colic pains also. Its seeds alone or in combination with sugar are used as a mouth freshener in India.

Distribution

It is widely cultivated in temperate and subtropical regions of the world for its aromatic fruits used as culinary spice. In India it is mainly cultivated in Rajasthan, Andhra Pradesh, Punjab, Madhya Pradesh, Uttar Pradesh, Gujarat, Karnataka, and Haryana states.

Description of Plant

It is a large perennial or biennial herb. Stem 2-4ft. high, hollow and cylindrical. Crop is protandrous and cross pollinated in nature. All parts of the plant are aromatic. Fennel seeds are exported annually. Fruit yield is volatile oil, of which the main constituents are *anethol*. The oil is used as flavouring agent. Various species of fennel are:

i. *Foeniculum vulgare* sub sp. *vulgare* (fennel, sweet fennel, wild fennel)

ii. *Foeniculum vulgare* sub sp. *vulgare 'Rubrum'* (bronze fennel)

iii. *Foeniculum vulgare* sub sp. *vulgare* var. *azoricum* (Florence fennel, bulb fennel, finocchio, anise)

iv. *Foeniculum vulgare* sub sp. *vulgare* var. *dulce* (sweet fennel, Roman fennel)

v. *Foeniculum vulgare* sub sp. *piperitum* (wild pepper fennel) Indian fennel is regarded as a distinct variety var. *panmorium*

Climate

It requires a fairy mild climate. Frost prone areas should be avoided as this crop is susceptible to frost. It is cultivated as *Kharif* as well as *Rabi* crop. Cool and dry climate is best for the cultivation of fennel crop. Dry and cool weather during the seed set increases seed yield as well as the quality of the produce.

Soil

Shallow sandy soils are not suited for fennel cultivation. Best soils for fennel cultivation are black cotton soil and loamy soil containing lime. Proper drainage is also an important requisite for commercial cultivation of fennel crop.

Cultivation

Kharif Fennel

Land preparation for nursery

The land should be brought to fine tilth by 2-3 ploughing. For transplanting one hectare land, nursery should be raised in 100 sq. meter area. Manures and fertilizers should be mixed in soil properly at the time of land preparation. Beds of 3 m × 1 m size should be prepared for nursery.

Seed rate and sowing

The recommended seed rate of *kharif* fennel is 2-2.5kg seeds/100m^2. Seeds should be treated with organo-mercurial fungicides like- Cerasan or Captan @3g/kg seed before sowing . The ideal time of sowing is 15th June. The time of transplanting is during August. About 45days old seedlings are used for transplanting at a spacing of 90 × 60cm or 100 × 60cm. In nursery, light irrigation should be given after sowing and second after 3 to 4 days after sowing and then light irrigation are given as and when required to keep the soil moist upto full germination. Tender seedlings should be protected from direct sunlight, till they are under danger, by erecting temporary shed over the bed. When seedlings are free from the danger of sunlight, temporary shed should be lifted.

Manure and fertilizers

In nursery, 250kg FYM, 0.6kg Nitrogen, 0.3kg Phosphorous should be applied per 100sq.m nursery area.

Table 4: Manure and Fertilizers

Manure and fertilizers	Dose	Time of Application
FYM	20t /ha	The whole quantity should be mixed in soil at the time of land preparation
Nitrogen	100kg/ha	40 kg of Nitrogen as basal dose. Remaining 60kg Nitrogen should be top dressed in two equal splits of 30kg each at an interval of 30 days and 60days at the time of earthing up after transplanting.
Phosphorous	60kg/ha	The whole quantity should be applied as basal dose.

Interculture

First operation/ hoeing at 20-25 days after transplanting. Earthing up are done after 60-75 days from transplanting to check the lodging of plants.

Weeding

Weed infestation is a serious problem in commercial cultivation of fennel crop. First hoeing and weeding is recommended at 30 days after sowing. Both mechanical and chemical control of weeds can be practiced. If herbicides are used for weed control, Pendimethalin is most effective. Pre-emergence application of Pendimethalin @1.0kg/ ha supplemented with one hand-weeding 50 days after sowing controls weeds effectively.

Irrigation

Fennel is a long duration crop, during dry spell in monsoon and after the monsoon; it should be irrigated as per the need. Generally 10-12 irrigations are given to fennel crop depending on soil and climate. Irrigation at the time of seed development is critical. First irrigation at the time of transplanting and subsequent irrigations should be given at an interval of 15-20days up to February.

Harvesting

It is done by cutting the umbels when 70 % umbels reach physiological maturity. After cutting the umbels they are dried in partial shade. While drying, care should be taken for better aeration and umbels should be turned if necessary. The seeds are separated by beating the dry umbels with sticks followed by winnowing. For seed purpose, fully matured umbels are harvested.

Yield

2000 to 2500kg/ha.

Varieties

Kharif fennel

Gujarat Fennel 1	Tall and bushy plants with oblong, medium-bold and dark green seeds; tolerant to sugary disease and leaf-spot; matures in 225 days; average yield is 16.5 q/ha; suitable for early-sowing; tolerant to drought.
Gujarat Fennel 2	Gives better yield than the old varieties. Time taken to reach maturity stage in 207 days.
RF 101	Tall, erect and with stout stem; long, and bold grains, matures in 150–160 days, average yield is 15.5 q/ha.
RF-35 or PF-35	Tall, spreading plants with medium-sized, hairless and green seeds; tolerant to sugary disease, leaf-spot and leaf-blight; matures in 225 days; average yield of 12.8 q/ha.

Plant Protection

These are similar in *Kharif* as well as *Rabi* Fennel.

Diseases

Damping off- In nursery, damping off seedlings is very common. It is a fungal disease. Infected plants show dehydration/wilting.

Control

1. Drenching of nursery beds with Blitox @ 20g per 10 litre water should be done to control this disease effectively.

Collar rot- It is common after transplanting. This disease appears in those areas where water stagnation near the plant stem. Collar portion of the plant start decaying and the plants turn to yellow colour and die later on.

Control

1. Disease can be controlled by spraying of 0.2% solution of copper fungicide and drenching by 1% Bordeaux mixture.

Ramularia blight- This disease is also very common after transplanting. At the beginning of this disease small ashy spots are seen on the leaves. Spots get enlarged and turn into brown colour. In severe cases the whole plant turns to brown colour, resulting in drying up of plant.

Control

1. In initial stage it can be controlled by spraying of 0.2% solution of Dithane M-45.

Pests

Aphid-After transplanting, aphid infestation is very common. They suck the saps of tender parts of plants and affect the growth, adversely.

Control

1. Spraying of 0.03% solution of Dimethoate or 0.05% solution of phosphamidone or any systemic pesticides.

Physiological Disorder

Fennel crop is vulnerable to frost damage which can be minimized by spraying 0.1% solution of sulphuric acid prior to the incidence of frost. Another disorder is Sugary Secretion. It is a physiological disorder, due to which flowers secrete sugary substance; attracting fungus from outside. Due to this disorder, quality and production of crop is affected. This phenomenon is observed in those plots which have been highly manured and irrigated under heavy dewfall and cold condition. Sugary secretion spreads on flower parts, stem etc. Due to this mould and other fungus develop on plants causing black and gummy appearance. The sugary secretion, attracts aphids too and these insects attack the plants. Combined effect of above gives false appearance of disease, but really sugary secretion is not a disease. For the control of this disorder, stop irrigation and manuring. Spraying of 0.03% solution of Dimethoate or 0.05 % solution of Phosphamidone or any systemic pesticide should be done to control the aphids.

Postharvest Management

Drying of Fennel Seeds

Harvested umbels are dried in sun for 1–2 days and then in shade for 8–10 days. Care must be taken not to sun-dry the umbels too long as long exposure of seeds to the sun reduces the quality of the produce.

Cleaning and Grading

After drying, fennel seeds are cleaned with the help of vacuum gravity separator or spiral gravity separator. Cleaned seeds are then graded based on its quality and then packed in jute bags for marketing.

Storage

Dried and clean seeds are stored in jute bags in damp-free aerated store.

3

Production Technology of Aromatic Grass

3.1 LEMON GRASS OR COCHIN GRASS

Botanical Name	*Cymbopogon* spp.
Family	Graminae
Origin	Native to South Asia, South-East Asia and Australia. East Indian Lemon Grass (*Cymbopogon flexuosus), also known as Malabar or Cochin grass is nati*ve to India, Sri Lanka, Burma and Thailand ; for the related West Indian lemon grass (*C. citratus*), a Malesian origin is generally assumed. Both the species are today cultivated throughout tropical Asia.
Chromosome number	2n=20,40
Commercial value	Part used in oil extraction is Leaf
Chemical constituents	The chief constituent of the oil is citral (80-90%). Lemon like flavour is due to citral
Research Institute	Aromatic And Medicinal Plants Research Station, Odakkali, Kerala

Importance

It helps to cope with fever, cough and cold, stress, high blood pressure, excessive fats in body. It lowers the cholesterol level, cleanse the body by eliminating toxic substances, improve the digestive system, improve blood circulation, cleanses other organs of our body including kidney, pancreas, liver, bladder etc.

and aid to deal with menstrual problems. The oil is distilled from leaves and flowering tops of Lemon grass. The oil has strong lemon-like odour, due to high percentage (over 75%) of citral in the oil. The characteristic smell of oil makes its use in scenting of soaps, detergents, insect repellent preparations. However, the major use of oil is as a source of citral, which goes in perfumery, cosmetics, beverages and is a starting material for manufacture of ion ones, which produces vitamin – A. The Citral rich oil has germicidal, medicinal and flavouring properties. An allied species called West Indian lemon grass (*C.citratus*) has low citral content in the oil and has meager trade in the country.

Distribution

In India lemon grass is widely cultivated in the states of Kerala, Karnataka, Tamil Nadu in the southern region, parts of Uttar Pradesh and Uttaranchal in the northern region and Assam in the north-eastern region. At present, East Indian lemon grass (*C.flexuosus)* is mainly cultivated in the western part of India.

Three species of lemon grass are generally identified.

i) *Cymbopogon flexuosus* (Nees ex Steud.) Wats., (2n=20,40).

It is known as *East Indian, Cochin or Malabar grass*. It is a tufted robust perennial grass, 2 m tall. It flowers freely. Leaves linear, lanceolate 125x1.7 cm. Panicles very large, drooping, lax, greyish-green, rarely with a purplish tinge, with the raceme pairs in dense masses, spreading, 100-135 cm long, slightly hairy, lower glumes of the sessile spikelets 3-4, 4-5 mm long, 1 mm wide with 1-3 definite or obscure intracranial nerves, shallowly concave with one to two depressions. *C. flexuosus* is commercially grown in Kerala and adjacent states. Its oil is popularly called as "Cochin oil". Under this species two varieties are identified based on the stem colour.

a) *C. flexuosus var. flexuosus*: The Red grass, locally known as "*Chumannapullu*". The stem is reddish or purplish in colour. It is recognized as the *true lemongrass* and is commercially cultivated. The oil yield is poor but the citral content of the oil is high (75-90%).

b) *C. flexuosus var. albescens*: The white grass, locally known as "*Vellapullu*", is characterized by the white colour of the stem. The plant is normally seen wild. The oil is poor in solubility. The oil yield is high but poor in citral content (60-75%).

ii) *Cymbopogon citratus* (DC.) Stapf. (2n=40,60)

It is known as *West Indian or American lemongrass. C. citratus* is a stemless perennial grass with numerous stiff tillers arising from short rhizomatous rootstock, making large tussocks. It seldom flowers in cultivation. Leaf blade narrow, linear, glaucous, drooping, 50-100 × 0.5-1.5 cm with scabrous margins. Ligule truncate, 0.2-0.8 cm long. Inflorescence rarely produced, a large loose panicle; spathe bracts long and narrow, sessile spikelets, awnless, linear-lanceolate.

iii) *Cymbopogon pendulus* (Nees ex Steud.) Wats.

It is known as *Jammu lemongrass*. It is white stemmed and dwarf in stature. The plant is frost resistant and suited to Sub-Himalayan areas of Northern India.

Description of plants

Lemon grass is a tall, perennial sedge throwing up dense fascicles of leaves from a short rhizome. The culm is stout, erect, upto 1.8 meter high. It reaches up to 2 m. tall under cultivation. Leaves are long, glaucous, green, linear tapering upwards and along the margins; ligule very short; sheaths terete, those of the barren shoots widened and tightly clasping at the base, others narrow and separating. It is a short day plant and produce profuse flowering in South India. The inflorescence is a long spike about one metre in length. It produces flowers and sets seed luxuriantly during autumn under short day condition only. Flowers borne on decompound spatheate ; panicles 30 to over 60 cm long.

Cultivation

Climate

It is very hardy, drought tolerant crop and grows under a variety of conditions. The most ideal conditions are a warm and humid climate with plenty of sunshine and required rain fall is 250-280 cm. per annum. The grass prefers an average temperature of 23-30° C. The plant is hardy and tolerant to drought. This crop is well suited for rainfed agriculture. It grows well at altitudes between 100 and 1200 m above MSL.

Soil

It is generally grown on poor soils along hill slopes, though it flourishes on a wide variety of soils ranging from rich loam to poor laterite. The grass grows best on well drained, acidic, loamy, medium to highly fertile soils under warm tropical climate with high well distributed rainfall is good condition for rainfed cultivation. The pH of the soil should be 4.5-7.5. It can tolerate alkalinity to some extent. In Kerala, it is grown as rain fed crop over sub marginal sandy, laterite soils and hilly slopes. However, it is cultivated throughout Andhra Pradesh either as a rainfed or irrigated crop in poor and marginal soils, wastelands and alkaline soils having pH 9.6.

Propagation

The crop is best propagated through seed raised in nurseries, 2.5 kg. of the fresh seed produces enough seedling for planting one ha. of land. It is also vegetatively propagated by splitting the clumps into slips. Clonal propagation is better as seed propagation leads to considerable genetic heterogeneity resulting in gradual deterioration of yield and oil quality.

Nursery raising

It is advantageous to raise the plantation through transplanting of seedlings whenever there is assured source of water. The transplanting of nursery raised seedlings is found to be superior to direct sowing of seeds. Fresh seeds are to be used for sowing as they are short viable and the viability is drastically reduced after six months of storage. Seed rate is 3-4 kg/ha for transplanting and 25-30 kg/ha for broadcast sowing. Seed method is popular in South India while in other states, rooted cutting obtained from slips. Seeds are grown in the nursery. Nursery bed should be well pulverized and make raised bed. The seeds are sown by hand on well prepared raised beds of 1m to 1.5m width in lines drawn at 10cm spacing at the onset of monsoon and are covered with a thin layer of soil. Seeds germinate in 5-7 days and seedlings are transplanted when 45-60 days old when they attain 12-15 cm height, they are ready for transplanting.

Rooted slips

For better quality and yield of oil it is recommended to grow lemon grass by slips obtained by dividing well-grown clumps. Tops of clumps are cut off within 20-25 cm of the root. The latter is divided into slips and the lower brown sheath is removed to expose young roots.

Land Preparation and Planting

Lemon grass is a perennial crop with a duration of 4 to 5 years. For better growth and establishment, initial land preparation is important. Two to three ploughing followed by making of ridges and furrows are necessary. Application of 10 kg Phorate or 8 to 10 kg neem cake at the time of last ploughing is recommended to control soil borne pests like nematodes, etc. Planting is done at the beginning of the rainy season. Seedlings are planted at a distance of

40×40 cm., 40×30 cm., 40×60 cm. apart depending upon fertility of land and inter-culture implements used. It is better to plant on ridges in areas receiving high rainfall. About 55,000 slips are required for one ha of planting. In case of rooted slips one or two slips are placed into each hole, about 15 cm deep at a spacing of 60×80 cm..

Manures and Fertilizers

Lemongrass is a soil exhausting crop. It is recommended to apply 30 kg nitrogen, 30 kg $P_2 O_5$ and 30 kg K_2O per ha basal dose at the time of planting. Remaining nitrogen (60 to 90 kg) can be applied as top-dressing in 3 to 4 split doses during the growing season. In soils having low fertility levels, the dose of nitrogen should be increased. In Zinc deficient soils 25 - 60 kg Zinc sulphate per ha. is applied. 100 kg.N /ha under Kerala condition. under North East 60 kg. N +50 kg P and 35 kg K per hectare are recommended (60 : 50 : 35 kg N : P : K /ha).

Irrigation

The improved varieties of lemon grass perform well with supplemental irrigation. lemongrass has a high water requirement. Where annual rainfall exceeds 650 mm, irrigation is not necessary. *C. citratus* is more drought tolerant than *C. flexuosus* overhead, flood and drip irrigation can be used. Depending upon the rainfall and its distribution the field is to be irrigated at an interval of 3 days during the first month and 7-10 day intervals subsequently. After the establishment of plants, irrigation schedule is adjusted depending on water holding capacity of the soil and weather conditions. Four to six irrigation would be critical to sustain the crop yield during summer months. Where rust is a problem, overhead irrigation might not be a good option.

Interculture Operations

The field is kept free of weeds for the first 3 - 4 months after plating. Similarly, weeding cum hoeing is done up to 1 month, after every harvest. Crop-weed competition period is the first 25-30 days after planting or harvest. Generally, 2-3 weeding are necessary during a year. Among herbicides Diuron at 1.5 kg/ ha and oxyfluorfen at 1.5 kg /ha are effective for weed control. Spraying Paraquat at 2-2.5 l/ha immediately after cutting the grass for distillation is an excellent method of weed control. Earthing up of plants after weeding and fertilizer application is beneficial as the roots have a tendency to grow above the soil.

Harvesting

Lemon grass flowers in winter season. The first cutting of grass is taken about 3 months after planting, thereafter every 6-8 weeks depending upon the fertility of the soil and other seasonal factors. Under normal conditions, three harvests are possible during the first year, and 3-4 in subsequent years, depending on the management practices followed. The grass is cut 10 cm. above ground level with the help of sickle and 5-6 cuttings can be taken in a year. Depending upon soil and climatic conditions, the plantation lasts on an average, for 4 to 5 years. The yield of oil is less during the first year but it increases in the second year and reaches a maximum in the third year; after this, the yield declines. On an average, 20 to 30 tonnes of fresh herbage is harvested per hectare per annum from 3-4 cuttings. The yield in terms of oil vary from 0.5% to 0.8% depending on the variety, season/month of harvest and age of the crop, with an average oil yield of 0.65%.

Oil yield

25 kg/ha in the 1st year and 80-100 kg/ha in 2nd year to 6th year (If well maintained).

Post Harvest Management

Drying

The grass is allowed to wilt for 24 hours before distillation as it reduces the moisture content by 30% and improves oil yield. The crop is chopped into small pieces before filling in the stills. It can be distilled in same distilleries as used for Japanese mint in India.

Distillation

Oil is obtained through steam distillation. The oil has a strong lemon like odour. The oil is yellowish in colour having 75-85% citral and small amount of other minor aroma compounds. The recovery of oil from the grass ranges from 0.5 - 0.8 per cent. It takes about 4 hours for complete recovery of the oil.

Purification of Oil

The insoluble particles present in the oil are removed by simple filtration method after mixing it with anhydrous sodium sulphate and keeping it overnight or for 4-5 hours. In case the colour of the oil changes due to rusting then it should be cleaned by steam rectification process.

Storage and Packing of Oil

The oil can be stored in glass bottles or containers made up of stainless steel or aluminium or galvanized iron, depending upon the quantity of oil to be stored. The oil should be filled up to the brim and the containers should be kept away from direct heat and sunlight in cool/shaded places.

In North India

Jammu Lemon grass (*C. pendulus*) is cultivated under irrigated condition as a source of citral. Cultural practices are similar. It is propagated by slips which are planted on flat beds. Spacing of 50 × 50cm is adopted. 260kg N, 80kg P_2O_5 and 120kg K/ha. NPK is recommended in 3-4 splits. Depending upon the planting period, 1-2 cuttings are taken in I year. II year onwards, 3-4 cuttings are available. Harvesting consists of fresh leaves and also dry or semi dried leaves at intervals of 60 days. The crop should not be allowed to flower profusely as it reduces the over all oil yield.

Rooted slip

Rooted cuttings are planted at 5-8 cm. depth. Cutting should not sow very deeply in rainy season otherwise cutting die. Cutting should be pressed in the soil.

Sowing time

In plains and subtropical area it is planted in July. If irrigation facilities are available. It may be planted in Feb.-March. Feb.-March planting gives 20% more yield in 1st year.

Weed management

Due to its planting in rainy season, crop severely suffered from weeds. 1.5 kg Diuron or 0.5 kg. Oxyflorofan herbicides used against the weeds.

Plant Protection

Pests

Stem-boring caterpillar (*Chilotrea* spp.)

It is the most important species infecting lemongrass. The caterpillar is white in colour with a black head and black spots on the body. It feeds on the stem. The first symptom of the attack is central leaf gets dried up and ultimately the whole shoot dies, resulting in a significant reduction in the yield of the grass.

Control

1. Burning of the dry stubble during the off-season in summer. The caterpillars lurking inside the stubble are therefore destroyed.
2. The affected shoots are pulled out and destroyed.
3. When attacks are severe, chemical control is needed.

Nematodes (*Tylenchorhynchus vulgaris* (Stunt), *Rotylenchulus renifomis* (Reniform), *Helicotylenchus* (Spiral) spp. *and Pratylenchus* (Lesion) spp.

It is soil borne and the whole plant is infected.

Control

1. Organic compost and mulching has advantages to incorporate soil organisms that are natural enemies of nematodes.
2. Soil solarization can be achieved by placing a clear sheet of plastic over the soil and leaving it for few days so that the heat of the sun kills the nematodes.
3. Crops such as *Tagetes* can also control nematodes.
4. Nematicidal chemical control should be used as last option owing to its toxicity. Application of Fenamiphos @11.2kg/ha is useful.

Diseases

Red Leaf Spot

C.O. *Colletotrichum graminicol*a

Brown spots with concentric rings in the centre appearing on the lower surface of the leaves. The spots may be formed on leaf sheaths and midrib. Later the spots merge to form bigger patches and the affected leaves dry away.

Control

1. Two sprays of Bavistin 0.1% just after the appearance of the disease at an interval of 20 days .
2. Three sprays of Dithane M-45 (0.2%) at an interval of 10-12 days

Leaf blight

C.O. *Curvularia andropogonis*

The disease appears in the form of minute, circular, reddish-brown spots mostly on the margins and tip of the leaves. These merge to form elongated reddish brown necrotic lesions, resulting in premature drying of leaves; older leaves are more susceptible to infection.

Control

1. Spraying Dithane Z-78 (0.2%) or 0.3% Copper oxychloride at an interval of 15 days.

Little Leaf or Grassy Shoot.

C.O. *Balansia sclerotic*

Stunted growth, little leaf formation in place of normal inflorescence.

Control

1. Spraying Dithane Z-78 (0.3%) just before flowering stage at an interval of 10-12 days.
2. Use of fresh seedlings for plantation and crop rotation.

Varieties

1. Sugandhi (OD 19) — It is adapted to a wide range of soil and climatic condition. A red stemmed variety with plant height 1 to 1.75 m and profuse tillering. The oil yield ranges from 80 to 100 kg per hectare with 85-88 per cent of total citral produced under rain-fed conditions (with life saving irrigation). Yield of herbage (t/ha/yr) is 50-55, Old yield (kg/ha/yr) 80-85.
2. Pragati — Developed by CIMAP (Central Institute of Medicinal and Aromatic Plants), Lucknow. It is a tall growing variety with dark purple leaf sheath suitable for north Indian Plains and *tarai* belt of subtropical and tropical climate. Average oil content is 0.63 per cent with 75-82 per cent citral.
3. Praman — Developed by CIMAP (Central Institute of Medicinal and Aromatic Plants), Lucknow. Evolved through clonal selection and belong to species *C. pendulus*. It is a medium sized variety with erect leaves and profuse tillering. The oil yield is high with 82 per cent citral.

4.	Jama Rosa	Very hardy with vigorous growth. The variety yields about 35 tonnes of herbage per ha. containing 0.4 % oil. The variety yields upto 300kg oil in 4-5 cuts in 16-18 months growing period.
5.	RRL 16	Average herbage yield of this variety is 15 to 20 tonnes/hectare/annum giving 100 to 110 kg oil. The oil content varies from 0.6 to 0.8 per cent (fresh weight basis) with 80 per cent citral. Yield of herbage (t/ha/yr) is 55-65, oil yield (kg/ha/yr) 120-150.
6.	CKP 25	A hybrid between *C.khasianum* × *C.pendulus.* Gives 60 t/ha herbage in North Indian plains under irrigation. The oil contains 82.85% citral.
7.	Kaveri (OD- 408)	It is white stemmed selection from OD-19 and is an improvement in yield in terms of oil and citral content. Kaveri needs high soil moisture to produce luxuriant growth and is evolved for river valley tracts.

	Varieties	Yield of herbage (t/ha/yr)	Oil yield (kg/ha/yr)
1.	O.D.-19	50-55	80-85
2.	S.D.-68	50-55	80-85
3.	RRL-16	55-65	120-150
4.	Peraman (Tetraploid mutant)	80-85	230 kg.

References

Post harvest Management of Lemon Grass.
http://www.nhb.gov.in/Horticulture%20Crops%5CLemongrass%5CLemongrass1.html
Varieties of Lemon grass
http://www.agri.puducherry.gov.in/lemongrass.html

3.2 CITRONELLA

Botanical Name	*Cymbopogon winterianus– Java citronella* *Cymbopogon nardus – Ceylon citronella*
Family	Graminae/Poaceae
Origin	Sri Lanka
Chromosome number	2n=20
Commercial value	Oil is extracted from leaves
Chemical constituents	The main constituents of oil are citronellol, geranial and citronellal.

Importance

Citronella is a tufted perennial grass, the leaves of which on distillation give an yellowish-brown essential oil with *citrus odour*. It is a perfumated perennial grass. It is used in perfumery and cosmetic industries. Citronella oil is used mostly in perfumery, both directly and indirectly. Soaps, soap flakes, detergents, household cleansers, technical products, insecticides, *etc* are often perfumed exclusively with this oil. It is also a valuable constituent in perfumery for soaps and detergents. It has a pleasant, sweet aroma like lemon. Citronella is occasionally used in traces in flower compositions of the citrus, cherry, ginger, etc. It is widely used ingredient in natural insect repellents (insecticidal properties). It has anti-bacterial, anti-fungal and antiviral properties. It has softening effect on the skin and also helps to combat excessive perspiration-making it extremely useful for use on feet. In trade, citronella oil is categorized into two types:

1. Ceylon citronella oil: obtained from *Cymbopogon nardus Rendle*, (inferior type).
2. Java citronella oil: obtained from *Cymbopogon winterianus Jowitt*, (superior type).

The Ceylon citronella oil was the more widely produced citronella oil, until the beginning of this century. However, the Java citronella oil slowly began to show dominance in the market because of its higher yield of oil. The chemical constituent of Java type Citronella is given below:

S.N.	Chemical	Percentage
1.	Citronellal	32-45%
2.	Geraniol	12-18%
3.	Geranyl acetate	3-8%
4.	Citronellyl acetate	2-4%
5.	Linalyl acetate	2%
6.	1-limonene	2-5%
7.	Caryophyllene	2.1%
8.	Linalool	1.5%
9.	Farnesol	0.6%
10.	Methylisoeugenol	2.3%

Distribution

The citronella grass has now been widely distributed throughout the tropics comprising Sri Lanka, Java, Central America, Guatemala, Taiwan, Brazil, East Africa, Congo, Malagasy Republic, The Seychelles, India and West Indies. In India it is cultivated in the States of Assam, West Bengal, Uttar Pradesh, Maharashtra, Karnataka, Tamil Nadu, Gujarat, Arunachal Pradesh, Manipur, Mizoram, Meghalaya, Nagaland and Tripura. The production of Java citronella far exceeds most other essential oils. The world production of citronella oil is about 7000 tonnes/year; the major producers being Taiwan, China, Indonesia and Guatemala. The chief importing countries are USA, UK, West Germany, Japan and Hong Kong.

Description of plants

It is a tufted aromatic perennial herb with shallowly rooted rhizome. Culms are tufted, robust, up to 2 m tall. Leaf sheaths are glabrous, reddish inside; leaf blades relatively thin, drooping 2/3 of their length, 40-80 cm tall, 1.5 to 2.5 cm wide, abaxial surface glaucous, adaxial surface light green, margins scabrid, base narrow, apex long acuminate. Spathate compound panicle large, up to 50 cm. Sessile spikelets are elliptic-lanceolate; lower glume, flat or slightly concave, 2-keeled, 3-veined between keels.

Two varieties are recognized in this species.

i) *Cymbopogon nardus* var. *lenabatu*

It is known as *Ceylon citronella.* It is the older form and less important than Java type. Ceylon type is cultivated in Sri Lanka. It is hardier and can be grown on poor soils. The leaves which are smaller than that of Java type yield an inferior oil with a lower citronellal content. The plant is a robust, stoloniferous perennial grass, 0.5-1.0 m high, leaves broad, panicles large and mostly compound, spikelets awnless, lanceolate and flattened on back.

ii) *Cymbopogon nardus* var. *mahapengiri*

This is sometimes recognized as a distinct species, *C. winterianus* Jowitt. It is known as *Java citronella.* Java citronella is a tufted aromatic perennial grass with fibrous roots. Plants are over 2 m in height. Clumps are stout, erect, terete, leafy and glabrous at the nodes. Leaf blades linear, with long acuminate tip and serrate margins. Ligules are scarious. Inflorescence is a very large compound panicle. Racemes 20 mm long, one sessile and the other pedicelled with the two lower spikelets, homogeneous, male or neuter, the remaining pairs in both racemes heterogeneous. Ovary superior with one carpel, unilocular. Stigma bifid, lateral and leathery. Ovule single and erect.

Climate

Warm humid climate where relative humidity is 70-80%, temperature 9-35° C are ideal for cultivation. It can tolerate higher temperature up to 45° C. Annual rainfall 200-250 cm. is required. Plant is not much pseophyte hence shade should be avoided. Although 180-120 m altitude is optimum, the plants are reported to grow well at the altitudes between 1000-1500 m.

Soil

The species has been found to grow well under varying soil conditions, but the sandy loam soil with abundant organic matter is the most suitable. Heavy clay soils and sandy soils do not support good growth of the plant. Rich sandy loam having pH of 5-6.5, or even more upto 8.5 pH is suitable. The crop is very sensitive to water logging hence drainage should be quite good.

Cultivation

Propagation

Citronella does not produce viable seeds, therefore, the species can be propagated only vegetatively by slips. Slips are obtained by dividing well grown clumps. Clumps are separated in a manner that each slip contains 1-3 tillers. Roots and leaves are trimmed off before planting. About 50 slips are obtained from an year old healthy clump.

Slip Preparation

Citronella is propagated by splitting the clumps of vigorously growing plant of 6-months to one year old into slips. The clump is gently dug out and separated into number of slips containing 2-3 tillers/slip. The fibrous roots and leaves are trimmed off at 25 – 30 cm height before planting. One-year-old clump on an average, gives about 50 slips.

Land Preparation

In the plains, it is cultivated in bed and channel system. Land should be well prepared. During preparation, 10-15t FYM per ha should be mixed with soil.

Planting Time

Although the plantation of Java citronella can be initiated anytime during the year, suitable planting time is the month of April-May or onset of monsoon. If irrigation facilities are available, then it may be planted in Feb-March under North Indian conditions. It produces 20-25% more yield in first year.

Planting Methods

The slips are removed from 1 or 2 year old plants and selected for planting in well prepared land at a spacing of 60×30 cm. Two slips are planted per hole to a depth at 15 cm. Slip requirement for planting is 55,000 slip/ha. If the planting is delayed, the slips may partially dry up resulting in poor plant population. The slips are planted vertically, about 10 cm deep. The planting should be done in such a way that the excess water is drained off quickly. This is because plants are extremely sensitive to temporary waterlogging, which adversely affects the growth of the plant. It is better to plant citronella on ridges to avoid waterlogging. The field should be irrigated immediately after planting if there are no rains within next 24 hours.

Manures and Fertilizers

Java citronella generally requires high dose of nitrogen for good growth. For medium and high fertile soils -250 kg N, 60 kg. P_2O_5 and 60 kg K_2O. For low fertile soils 300 kg N, 60 kg P_2O_5, 60 kg K_2O/ha. ¼ Nitrogen and full dose of P and K at the time of planting and remaining Nitrogen should be after every cutting should be given. Plant also responds micronutrients like Fe and Zn. The crop sometimes may show yellowing of growing leaves indicating a mixed deficiency of Zinc, Molybdenum and Sulphur. Two or three foliar sprays of a complete micro-nutrient mixture at fortnightly intervals is sufficient to combat the deficiency.

Irrigation

For proper development of the plant, moisture should be retained in the soil. Java citronella requires sufficient moisture for good growth and yield of the leaves. In the areas where annual rainfall is about 200-250cm, well distributed over the year and the humidity is high, supplemental irrigation is not necessary. First irrigation after planting and subsequent irrigation is given at every after cutting. In drier months, however, irrigation is provided twice a week during the first month of planting and thereafter once in 5 days. Proper drainage facilities should be present in the soil.

Intercultural Operations

The plantations should be kept weed free. When the plants have established themselves and formed bushes, the problem is not so severe because of the very nature of growth of the bushes. The bushes do not allow weeds to grow around them by cutting off the sunlight. However, in the newly established plantations and after each harvest, the weeds grow in the inter-row spaces and weeding is essential. Weeding and hoeing is required at 15-20 days after planting. After every cutting, weeding is necessary. Use Diuron 1 kg (a.i.) in 500-600 liter of water at the time of planting and after first cutting. Second year before monsoon earthling should be done. It checks the stagnation of water. It is recommended that the plantation should be uprooted after 3-4 years and rotated with some small legume species. Horsegram is a very good rotational crop in south, cowpea or sunhemp (*Crotalaria species*) is recommended for north Indian plains.

Harvesting

Citronella is cultivated for essential oil. Although, all the plant parts contain oil, leaves contain the maximum amount of oil. Therefore, only the leaves should be harvested. Harvesting is done by sharp sickle at about 20-45cm above the ground. The number of harvests, which can be taken during a year, depends upon the growth of the plants. First cutting is ready in 4-5 months after planting and there after cuttings are taken at two months interval before winter period. The leaves are ready for first harvest, about 6 months after planting. The second and subsequent harvests can be taken thereafter at 2.5-3 months interval. Harvesting too soon and too late affects the quality of oil adversely. The delay also causes the leaves to dry up resulting in decrease in yield of oil. While harvesting, only the leaf blade should be cut and the sheath should be left. This is because the sheath contains only little and poor quality oil. Flowering should be discouraged as it causes aging in plants and reduces their life span. The

Java citronella plantations remain productive for 5-6 years but the yield of leaves and oil is highest during second and third years, after which it starts decreasing.

Yield

On an average, the oil content is about 1% on the basis of fresh weight of leaves. Year wise details of yield is given below.

Year	Oil yield (kg/ha)	Herbage yield (tonnes)
I	140-150	150-200
II to III	200-300	250-300
IV	200	-
V	100	-

Plant Protection

Diseases

a) Leaf blight or *Curbularia* leaf blight

It is the most important disease of this crop caused by ***Curbularia eragrostidis.***

Control

1. Spray of Carbendazim (Bavistin at 1.0g/liter) followed by Benomyl of Benzimindazole group (Benlate 50 wp) @0.2% at 10 days interval.

Insect and pest

a) Mites

During dry season the crop may be affected by yellowing and crinkling of leaves due to mites. The emerging leaves become wrinkled and pale. It may fail to open and the part may die.

Control

1. Spraying a mixture of Docofol (Kelthane) @ 2 ml/lit and Ethion (Tafethion, 50 EC) @ 1 ml/lit along with a wetting agent like Triton AE or Enditron AE @ 0.5 ml/lit.

b) Stem borer

Control

1. These can be controlled by applying Furadan @ 20 kg/ha or by spraying 0.5 % Dimethoate or 400 to 500 kg Neem oil cake.

Varieties

1. Mandakini	Clonal selection, gives a little less herb yield (35 t/ha) and oils (118 kg/ha). The variey is suitable for hills and tarai tracts of Himalayas (CIMAP, Lucknow).
2. Manjusha	Clonal selection, gives a herbage yield of 43 tonnes/ha. and 150 kg/ha of oil per annum. The variety is suitable for Indo-Gangetic plains (CIMAP, Lucknow).
3. Manjiri	This variety has been released by University of Agriculture Sciences, Bangalore. It is an elite mutant clone of Manjusha M3-8. It has been found to possess 50-90% more oil, high cirtonellol and low elemol content. It has profuse tillering and rapid growing ability thus producing a high herb yields.
4. CIMAP Bio-13	Released from CIMAP, Lucknow and developed from tissue culture technique.
5. Java-2 and Jorhat-2	HYV developed from CIMAP for southern and eastern states of India.

3.3 PALMAROSE OR RUSA OR RUSHA GRASS

Botanical Name	*Cymbopogon martini* var. *Motia Cymbopogon nardus* – Ceylon Grass.
Family	Graminae
Origin	India
Chromosome number	2n=40
Commercial value	Essential Oil which is distributed in all parts of the grass *viz.*, the flower-heads (containing major portion), leaves and stem.
Chemical constituents	Myrcene, linalool, geraniol, geranyl acetate, dipentene and limonene.

Importance

Palmarosa is a tall perennial grass, the flowering tops and foliage of which contain a sweet-smelling oil of *Rose-like odour*. Two types of palmarosa grasses have been recognized, *viz*, *Rosha* or *Russo grass* and *Gingergrass*. Rosha grass yields a superior oil which is used in perfumery, particularly for flavouring tobacco and blending soaps due to the lasting rose note it imparts to the blend. Gingergrass oil is poor in quality and is mainly used as a cheap perfume in countries bordering the Red Sea. Oil of palmarosa contains geraniol. The oil is useful in imparting rose like aroma to a wide variety of soaps, tobacco products etc. The oil of palmarosa is commercially preferred to ginger oil. Oil has antiseptic, antiviral and bactericidal properties. Palmarosa oil calms the mind, yet has an uplifting effect, while clearing muddled thinking. It is used to counter physical and nervous exhaustion, stress-related problems and nervousness. It is most useful during convalescence and cools the body of fever, while aiding the digestive system, helping to clear intestinal infection, digestive atonia and anorexia nervosa. It is effective in relieving sore, stiff muscles. Palmarosa oil moisturizes the skin, while balancing the hydration levels and stimulating cell regeneration. It balances production of sebum, to keep the skin supple and elastic and is valuable for use with acne, dermatitis, preventing scarring, rejuvenating and regenerating the skin, as well as fighting minor skin infections, sore tired feet and athlete's foot.

Distribution

The exact location of origin of the crop is not clearly known. It grows wild in India, particularly in Madhya Pradesh, Maharashtra, Jodhpur (Rajasthan), Andhra Pradesh, Karnataka, Uttar Pradesh and Orissa. Palmarosa oil has been distilled in India for more than 50 years, mostly from wild growths. The plant was successfully introduced into Java, The Seychelles, Guatemala and Brazil where commercial production was undertaken to a limited extent. About 60-70 tonnes of palmarosa oil is produced in India annually both from cultivated and natural sources.

Description of Plant

It is a drought hardy grass attaining a height of 1.5-2.5 m having hairy and fibrous shallow root system with long linear lanceolate leaves. This wild growing plant has long slender stems and terminal flowering tops. The grassy leaves from which the essential oil is derived is extremely fragrant. It produces large fawn coloured inflorescence while, hairy star like spiked flowers.

In this species, the following two varieties are recognized which are morphologically very similar but differ considerably in their habitat and also in the essential oils.

i) *Cymbopogon martinii* var. *motia* (2n=40)

It is known as *Palmarosa, Rosha or Russa grass*. Its leaves are lanceolate, 50 cm long, 1-3 cm broad. Panicles 10-30 cm long, turning reddish, often very bright when mature. Racemes 15-19 mm long, in pairs, each consisting of many pairs of spikelets. In each pair, one is sessile and hermaphrodite and the other is pedicelled and male. Sessile spikelets 3.5 mm long. Glumes 3.3-4.0 mm long. Awns 11.4-14.0 mm long. Lower glume of fertile spikelet deeply channeled.

ii) *C. martinii* var. *sofia* (2n=20)

It is known as *Gingergrass*. Morphologically it resembles motia grass, but the oil is inferior. *Jamrosa* is an interspecific hybrid between *Cymbopogon nardus* var. *confertiflorus* and *C. jwarancusa*. Its oil contains 75-80% geraniol and 15-20% geranyl acetate. The oil has more terpenic, limy or grassy by-odours in contrast to the sweet, rosy odour of palmarosa oil.

Soil

It requires wide range of soils, which are free from water logging, but a well-drained loamy soil with pH of 6 to 7 is best. With available irrigation facilities it can be grown in poor sandy to heavy fertile soils and also in marginal and sub marginal lands.

Climate

It is a hardy plant can grow in varying altitude right from sea level. It stands well in places receiving rain fall from 75 to 150 cm.and temperature 10 to 36°C. But it does not with stand stagnant water. It requires exposed sunlight and does not perform well under shady situations. It needs an annual rainfall of 150 cm.

Cultivation

Preparation of Land

Field is prepared before the onset of the monsoon. It is ploughed and harrowed to a fine tilth. All the stubble and weeds are removed. At the time of last ploughing FYM @10t/ha is incorporated into the soil. Nursery raised bed should be prepared by a well pulverized of the mixture of FYM and soil. Best time for sowing is April-September. Seedling is ready for transplanting when they are attains a height of 15 cm.

Spacing

Transplanting of nursery –raised seedlings are done at a spacing of 90 × 60 cm. Closer plants spacing of 30 cm × 30 cm improves oil production. In worth, it is recommended as a mixed crop along with Basil to get higher profit from unit area.

Propagation: Raising Seedlings in Nursery

1. Nursery raised seedlings: Nursery beds are well prepared by end of April to mid May under partial shade. Raised beds are preferably as the seeds are not washed off by irrigation. Beds are made by adding enough FYM, compost or organic matters and channels are dug for irrigation. The seeds are sown in lines, 15-20 cm apart. About 2- 2.5 kg seeds per 100m^2 are adequate to give seedlings for planting. As the seeds are small and light, they are mixed with fine soil for obtaining even distribution and ease in sowing. Germination starts within 2 weeks and seedlings are ready for transplanting within 3-4 weeks. About 0.2-0.5% solution of urea may be sprayed for good vegetative growth. These seedlings are transplanted in June and July. The healthy and established seedlings which are 15 cm high are planted 60 cm apart.
2. Root cutting/Slips: For obtaining better quality and yield of oil it is recommended to grow Palmarosa by slips obtained by dividing well grown clumps. Tops of clumps are cut off within 20-25 cm of the root. The latter is divided into slips and the lower brown sheath is removed to expose young roots. Plants giving good yield and high quality oil are to be used. However rate of establishment of root slips is very poor as compared to nursery transplants. Slips can be planted in June-July or rainy seasons. It is best to raised by transplanting the nursery-raised seedlings or by root cuttings of healthy plants.

Sowing

First prepared the raised nursery bed and seeds are mixed with fine soil in the ratio of 1; 10 for even distribution and ease in sowing. Seeds are sown in lines at 15-20cm apart. About 2.5kg seeds are adequate to give enough seedlings for planting one hectare. The beds are watered lightly and regularly. Seedling ready for transplanting when they are 15 cm high. It takes 3 to 4 weeks from the sowing. Slips are planted in June –July .The rate of establishment of rooted slips are very poor as compared to nursery transplant, hence seed propagation is commercially utilized.

Manure and Fertilizer

As palmarosa is a long duration crop and removes substantial amount of nutrients from the soil for producing herbage, it is necessary to replenish the soil with rich soils. About 10t/ha FYM, 40 kg N, 50 kg P_2O_5, and 40 kg K_2O as a basal dose is recommended. about 60 kgN/ha is applied in three split doses during the growing season. The application of NPK should be repeated in subsequent years. Manuring may not be required for the first two years. By manuring rich soils, the vegetative growth is increased and oil content is reduced. In North–eastern regions of India 60:40:40 kg NPK/ha should be applied. Application of micronutrients like iron (as $FeSO_4$) and manganese (as $MnSO_4$) improves the growth, herbage and oil yield.

Irrigation

It depends upon the climatic conditions. palmarosa requires irrigation fortnightly during the growing season for better growth and recovery of essential oils. two irrigations in rainfed crop is sufficient to increase the herbage and oil yield.

Weeding

Regular weeding especially of the weeds which imparts their own odour is necessary to maintain the quality of oil.

Harvesting

The essential oil is distributed in all parts of the grass, *viz.* the flower-heads (containing major portion), leaves and stem. Recent studies at NBPGR, Delhi shows that oil from the whole plant is of good quality and economical to produce as per table given below:

SN	Plant part	Essential oil (fresh weight)	Geraniol (%)
1.	Whole plant	0.53	72.4-86.5
2.	Inflorescence	0.83	72.5-84.5
3.	Leaves	0.58	92.9-94.6
4.	Stalk	0.01	-

It is recommended to harvest the crop 7-10 days after opening of flowers. The right time for harvesting is when leaves contain high oil % just before blooming period. Besides superior quality of oil with roseous green odour is obtained at an early seed setting stage rather than full bloom stage when the oil yield is slightly higher. First harvesting starts after 6 months of transplanting and in full bloom. The number of harvest depends upon the climatic condition of the place of cultivation and method of crop management. During the first year, usually one crop is obtained in October-November, whereas 2-3 crops are obtained in the subsequent years in subtropical areas in the North Indian plains. Four harvests are taken in tropical areas of South and North-East.

Usually the grass is cut at a height of 5-8cm from the ground level and the whole plant is used for distillation. The cut grass is allowed to wilt in shade for about 24-48 hours before distillation. The plantation can be maintained for about 8 years to 10 year, but oil yield starts declining from the 5th year. Superior quality of oil with resource green odour is obtained at in early seed setting stage rather at full bloom stags when the oil yield is slightly higher. The maximum yield of oil is obtained when the entire plant is at a full flowering stage. The harvested herbage is spread in the field for 4-6 hours to reduce its moisture by 50% and such semi-dry produce can be stacked in shady cool space for few days without much loss of its oil.

Yield

Palmarosa plantation remains productive for about eight years. However, the yield of grass and oil starts decreasing from the fourth year onwards. It is, therefore, recommended that the plantation be kept only for four years. Normally 200-250 q/ha of fresh herbage is obtained in first cutting and between 250-320q/ha in second and subsequent harvests upto three years under irrigated conditions. On an average, 200kgs of oil are received during the growing period of 15-16 months.

The yield of oil for the first four years is as under :

1st year 60kg/ha

2nd year 80kg/ha

3rd year 80kg/ha

4th year 80kg/ha

15-20 tonnes of herbage/ ha in a year and Oil yield is 50-60 kg /year.

Plant Protection

No serious insect-pests has been reported under this crop.

Disease

1. Red Leaf spot

CO- *Colletotrichum graminicola*

This disease is much ubiquitous in all the Palmarosa growing regions during rainy season. Under warm and humid conditions, the disease spreads rapidly and may cause upto 35% reduction in oil production. The disease perpetuate in all the left out stumps and stubbles on the field. Small brown spots are scattered all over the ventral portions of the leaves. It also affects leaf sheath and mid ribs.

Control

1. Two sprays of bavistin (0.1%) just after first appearance of the disease at an interval of 20 days along with three sprays of Dithane M-45 (0.3%) at an interval of 10-20 days controls disease.

2. Little leaf or grassy shoot

CO- *Balansia solerotia*

In place of normal inflorescence little leaf formation occurs, plants are also stunted. Oil yield and seed production are adversely affected.

Control

Spray Dithane Z-78 (0.3%) before flowering stage at an interval of 10-12 days.

3. Smut

CO- *Tolyposprium christenseni*

This is an inflorescence borne disease that adversely affects the oil yield and seed production. The symptoms are characterized by the transformation of seed into the smut sori covered by dull yellow membrane.

Control

Seed treatment with organomercurial fungicides before sowing in the nursery. Spraying of Dithane Z-78 (0.3%) at the time of flower initiation.

Varieties

S.N.	Varieties	Description
1.	Trishna	It is a high yielding synthetic variety developed by poly cross progenies of four elite inbred by CIMAP Lucknow. It possesses straw yellow clumps, long compact inflorescence and has profuse tillering. This improved cultivar registers 40% more oil yield and its geraniol content is as high as 93%.
2.	IW -31245	Profuse tillering 20-25t/ha of herbage per cutting under low irrigation. Oil yield 150kgs in cuts during 16-18 months. Oil contains 90% total gerniol. It has superior rosaseous green odour.
3.	Jam Rosa (RRL-82)	It is a hybrid between *C. nardus* var. *confertiflorus* × *C. jawarancusa* developed from Regional Research Laboratory (RRL) Jammu. Plants are bushy with small internode and more number of tillers per clump. It has vigorous growth high yield of 35-45 t/ha of herbage. It yields between 0.3-0.4% of oil in the herb which is rich in geraniol content (54-83%) and gernail acetate (17.89%). The oil yield was estimated between 130-150 kg/ha in the second year.
4.	IW 31244 and RRL,	Superior selections identified by NBPGR

3.4 VETIVER (*KHUS*)

Botanical Name	*Vetivaria zizanioides* (Linn.) Nash
Family	Graminae
Origin	India
Chromosome number	2n=20
Commercial value	Roots
Chemical constituents	Vetiselinenol and Khusimol

Importance

It was earlier named as *Andropogon muricatus* Retz. Later it was recognized as *Vetiveria zizanioides* (L.) Nash. Roots are used to make aromatic mats, brooms, screens, pillow and mattresses along with bamboo splits. Roots are used in pharmaceutical preparations. Widely used in aromatherapy and perfume, *Khas-Khas* is a cooling agent, tonic and blood purifier. It is used to treat many skin disorders and is known to have a calming effect on the nervous system. Its other medicinal uses include ringworm treatment, indigestion and loss of appetite, treat gastrointestinal disorders like flatulence and indigestion. It is a natural cleanser that is used in soaps, shampoos and perfumes. Oil is consumed in soap making. The roots are also used in pharmaceutical preparations both in the allopathic and Ayurvedic systems of medicines. It is used as a stimulants, refrigerants and stomachic.

Its main chemical components are benzoic acid, vetiverol, furfurol, a and b-vetivone, vetivene and vetivenyl vetivenate. The chemical components of the oil obtained from the plant is benzoic acid, furfurol, vetivene, vetivenyl vetivenate, terpinen-4-ol, 5-epiprezizane, Khusimene, a-muurolene, Khusimone, Calacorene, β-humulene, a-longipinene, selinene, d-selinene,d-cadinene, valencene, Calarene,-gurjunene, a-amorphene, Epizizanal, 3-epizizanol, Khusimol, Iso-khusimol, Valerenol, β-vetivone, α-vetivone, vetivazulene.

Distribution

This grass is native of India and is found through out the plains and lower hills (upto 1220 meters in altitude) of India, Burma and Sri Lanka. It has been widely distributed in the tropics. Prior to second world war, Java was the principal producer of vetiver oil, but Haiti, Reunion and India are now the main producers, with subsidiary production in Congo, Angola, Brazil and Guatemala. The world production of essential oil is around 300 tonnes/annum. The crop is also cultivated in Indonesia, Malaysia, Philippines, Japan, Angola, Belgian Congo, Dominican

Republic, Argentina, British Guiana, Jamaica, Mauritius and Honduras. In India it is cultivated in Rajasthan, Uttar Pradesh, Karnataka, Tamil Nadu, Kerala and Andhra Pradesh.

Description of Plants

A dense, clumping perennial grass, to 1.5 m in height, native in India and Ceylon. 'Monto' is a sterile (non-seed producing) variety specially selected not to become weedy. In its natural environment, vetiver grows on riverbanks up to an altitude of 600m. It requires a hot and humid climate. It is adaptable to a wide range of soil and climatic conditions. It can be established on very acid, sodic, alkaline or saline soils. Vetiver tolerates very high levels of aluminium, manganese and a range of heavy metals in the soil. Due to its extensive and deep root system, vetiver is very tolerant of drought. It can stand extreme heat (50°C) and frost (-10°C) and can be established in areas with an annual rainfall from 450 mm and higher. Vetiver is sensitive to shade and this will slow growth, especially in young plants.

Climate

It prefers a mild climate but can be grown both in wet and dry or arid tropical conditions. Under temperate or warm winter hills areas, its growth remains stunted. It prefers temperature of 25°-30° C and relative humidity- 70-80%. Shade affects the unfavorable development of root system.

Soil

The plant is sufficiently hardy and grows on almost all types of soils. Light soils, however, should be avoided as the roots obtained produces very low percentage of oil. Red lateritic soils with abundant organic matter are considered ideal as the roots produced in such soils are thick and contain more essential oil. Heavy soils make harvesting of the roots difficult, with a loss of the finer roots which contain most of the oil. Loose sandy soil, preferably on the sloppy hills. In such soils only, the roots can be easily pulled without much loss of thin roots. It can be grown even on saline and alkaline soils with a pH range of 8.5-10.0.

Propagating Material

The grass is capable of both vegetative and sexual reproduction in nature. It is propagated through tillers and slips. Tillers take more time so slips are better than tillers, The top of the slips are cut down before planting to prune. It prevents transpiration losses and gives better survivability. Slips are planted in pits, about 5-8 cm deep made with a pointed stick. 2-3 slips are put in each pit.

Planting System

Vetiver should not be planted in shaded areas because it affects root development. Different planting systems are adopted while planting and they are as follows:

There are three different systems of planting adopted by different growers.

1. System I: Conical ridges, 30-38 cm high and 48 cm apart are made at the summit and the slips planted 23 cm apart on the summit.
2. System II: The land is laid out into beds of 30 cm high, 68 cm wide and 45 cms apart edge to edge and the slips are planted on these in two rows 22.5 cm apart, leaving 22.5 cm on either sides.
3. System III: The beds are made 45 cm high, 60 cm wide and 30 cm apart edge to edge and two rows, 30 cm apart, are planted on these leaving 15 cm on either side. The spacing within the row is also 30 cm in this system.

Vetiver slips 10 days after sowing

The slips are planted in pits, 5-8 cm deep. Two to three slips are planted in a row. 1,50,000 – 2,25,000 slips per hectare with 2-3 in one of hole is recommended with II system of planting . Planting Time is from February to October. but the best time to get higher oil under South Indian conditions is June- July.

Manures and Fertilizers

Ground nut cake or cattle manure or wood has beneficial effect on root yield of vetiver. Application of 22.5: 22.5: 20 kg NPK/ha provide good root yield.

Important Operation

For good growth in first year topping operations should be done. 2nd cutting before flowering increases oil yield and 3rd cutting 1 month before harvest also increases yield.

Harvesting

Roots are harvested after 15-24 months of planting. Harvesting can be done even earlier. Although the young roots give a higher yield of oil, this will be of low specific gravity and lacking in the valuable high boiling constituents. For good percent of oil the root should not be touched before two years. Best time for harvesting is in the months of April-May. Harvesting start with a stem portion and it is cut at a height of 15-20 cm and the clumps are uprooted then 50-60% roots come away with clumps. Young roots are tender, thin, and almost hairy like on pulling they break easily and stay in ground. The odour of then light oil is 'green earthy'. Older well developed thicker root yield an oil of better quality with better odour and darker of colour than younger roots. If the roots stay in the ground for over two years, the yield of oil diminishes considerably and the oil becomes very viscous with a dark colour but of a high quality. The length of the roots varies from 10-35 cm according to the condition of growth, soil, climate, etc of the locality. The roots are separated from the plants, washed to remove the adhering soil and dried under shade for 1-2 days which improves the olfactory quality of the essential oil. Prolonged drying in the sun reduces the oil yield.

Vetiver oil is extracted by water distillation and steam distillation of the roots. In both methods, roots are first cleaned, kept in water for about 12 hours, chopped into small pieces (10 cms) then distilled. Steam distillation is more economical, yet it is not preferred by majority of the commercial distillation units because oil produce by steam distillation does not give the same colour as that by water distillation. Distillation time varied from 12-36 hrs.

Yield

5 to 7 tonnes of roots per hectare and 15 to 16 kg oil.

Varieties

Basically there are two types of varieties (1) Seeding type and (2) Non seeding type.

1. **Seeding type:** These types grows wild in North India. The oil of some of the North Indian types *viz.,* Bharatpur, Akila and Musanagar strains generally have an aroma superior to that of oil derived from South Indian grasses.
2. **Non seeding type:** These type of varieties are grown in South India.

The Indian Agricultural Research Institute, New Delhi have developed superior hybrids *viz.*, Hybrid 8 and Hybrid 7 which are found to be superior in terms of root and oil yield. Hybrid 7 has found to perform better in Tamil Nadu conditions whereas, Hybrid ODVI 3 was identified as a superior variety than local Nilambur type in Kerala.

Other varieties are Dharini, Gulabi, H-9 and Kesari.

3.5 GERANIUM

Botanical Name	*Pilargonium graveolens* H.
Family	Geraniaceae
Origin	Cape Province (South Africa)
Chromosome number	2n=88
Commercial value	Leaves
Chemical constituents	β-citronellol (21.9%), citronellyl formate (13.2%), geraniol (11.1%), 10-epi-γ-eudesmol (7.9%), geranyl formate (6.2%) and (l)-linalool (5.6%).

Importance

Geranium oil is almost used in cosmetics and perfumery. It has antiseptic properties; astringent qualities. It helps in relieving symptoms of anxiety and menopausal symptoms. It helps in healing bruises, cut and scrapes, eczema, nail fungus and sun burns as well as helps with insomnia. It is a great insect repellent and is well known in the ancient world as a tick repellent for dogs and human alike.

Distribution

Even though, Cape Province in South Africa is native place of Geranium where it grows in a state of nature, but do not enjoy the position of a producer of its volatile oil. It is grown in France, Belgium, China, Spain, Morocco, Madagascar, Reunion Islands, Egypt, Congo, Russia and India for the production of geranium oil. In India, Maharashtra, Andhra Pradesh, Tamil Nadu, Jammu and Kashmir, Himachal Pradesh and Kerala are suitable for its cultivation. The major producers of geranium oil are Reunion Islands, Egypt, China, Algeria, Morocco, India and Russia.

Description of Plants

It is a bushy aromatic perennial shrub with cylindrical stem. Leaves simple, alternate, stipulate, broadly cordate with palmatisect primary lobes and pinnatisect secondary lobes, pubescent and aromatic; inflorescence umbellate, bracteate; flowers pentamerous, basically bisexual, hypogynous; calyx free quinquincial; corolla pink, zygomorphic, polypetalous; stamens 10; filaments subequal, united at the base; ovary superior, pentacarpellary, syncarpous, hairy; stigma 5 lobed.

Cultivation

Climate

It can be grown in temperate, subtropical and tropical climate at various altitude from 1000-2000m but a Mediterranean type of mild climate with a low humidity, warm winter, mild summer temperature and an annual rainfall of 1000-1500 mm is ideal for the crop. It thrives best in sub tropical climate with a temperature ranging from 5-23°C. Temperature below 3°C kills the plants. Geranium flourishes well where rainfall is distributed throughout the year but water logging may causes or results in root rot and stunting growth. Plant is quite resistant to drought but not for frost.

Soil

It is a shallow rooted crop hence required a well drained soil. The crop is found to perform well in the peaty and latertic soil with pH 5.5-7, though a calcium rich porous soil is much desired. Drainage in the soil must be good. Saline, alkaline and damp soils are unsuitable

Propagation

There is no seed setting in geranium. The plant is propagated by stem and root cuttings. Terminal cuttings root earlier than middle and basal cuttings. Terminal cuttings of 10-15 cm long are the best suited with 3-4 buds. Middle cutting and basal cutting can also induced to root up to 80% and 65% respectively by treating with growth regulator like IBA or IAA at 200ppm for 6 minutes.

Planting

For planting 25000-30000/ha cuttings are required. Rooted cuttings are raised in the nursery during September-October. If geranium is planted in winter season, then directly cuttings are planted in the field. Close spacing of 8-10 cm is kept in nursery. Cut ends of stem cuttings are dipped in 3% solution of wet ceresin. The cutting is ready for transplanting after 2 months at 60 × 40 cm spacing in the main field.

Fertilizers and Manures

Application of well decomposed 10-12 tonnes/ha of FYM, 60 kg Single Super Phosphate and 40 kg K_2O/ha at the time of planting. While, N is applied upto 200 kg/ha/year in six equal splits to cover the 3 harvests. 50 kg N per ha is applied at 20-25 days after transplanting. A second dose of nitrogen at 35 kg/ha is applied about two months after first application. Third dose of nitrogen (35kg/ha) is given at first harvest and the fourth at the second month after first harvest (35kg/ha). In addition, application of micronutrients of 20kg/ha of zinc sulphate and 10kg/ha of boron has been reported to increase the herbage yield. Application of copper @20kg/ha and molybdenum@30kg/ha/year in four split doses after each harvest has been found to increase the yield.

Irrigation and interculture operations

Irrigation is provided daily for the first 3-4 days, on alternate days till two weeks and weekly thereafter. The crop requires weeding 20 and 40 days after planting and hoeing after harvest.

Harvesting and Yield

It matures in 100-120 days when leaves begin to turn light green and exhibit a change from lemon like odour to that of rose. The green leafy shoots are harvested with a sharp sickle. Three harvests can be taken per annum and the crop remains in the field for 4-6 years. The yield of fresh herbage/ha/year from the 3 harvests is about 15000-18000 kg leaves can be harvested from a hectare in a year and they may yield 15 kg oil on distillation.

Plant protection

Pests

1. **White ants**- It attacks on the roof of the plant.

Control

(i) Use Heptachlora @25kg/ha. in soil or Chloropyriphos @ 800 ml per ha.

2. **Root-knot nematodes *Meloidogyne incognita* and *M. hapla***

Control

(i) Application of Aldicarb at 20 kg/ha is effective.

Disease

1. **Stem rot or Root rot**

Control

(i) After planting of rooted cutting apply Bavistin 1g or dipping of cutting in chlorothelonil @ 0.3%.

2. **Wilt disease:** It is caused by *Fusarium oxysporum* and *Botryodiplodia theobromae.*

Control

(i) Spray of systemic fungicide benomyl is effective.

Varieties

Algirian, Reunion, PG-20, PG-7, Sel-8, Hemanti, Bipuli, Kunti.

3.6 DAVANA

Botanical Name	*Artemisia pallens* Wall.*ex* DC.,
Family	Asteracae
Origin	Himalayan Regions. Southern part of India, especially to the states of Karnataka, Tamil Nadu, Andhra Pradesh and in Maharashtra
Chromosome number	2n=16
Commercial value	Flowers
Chemical constituents	Devanone

Importance

Davana is a delicate, erect, branched aromatic herbaceous plant, the flowering top of which yields an essential oil which is extensively used in high grade fine perfumes. The oil is used for flavouring cakes, pastries, tobacco, beverages, sausages and preserved products. Leaves are very small, bluishgreen with yellow flowers and inconspicuous. They are highly valued in the making floral decorations garlands and bouquets. The leaves and flowers are also utilized in traditional Ayurvedic medicinal formulations. Used in Indian folk Medicine for the treatment of diabetes mellitus led to significant blood glucose lowering effect in glucose-fed hyperglycemic and alloxan-induced diabetic rate. Essential oil of davana is useful as antiseptic and disinfectant.

Distribution

Davana is probably a native of South India. Karnataka, Tamil Nadu, Uttar Pradesh and Andhra Pradesh are the major producing states.

Description of Plants

It is an erect branched annual herb 45-60 cm tall and covered with greyish white tomentum. Leaves alternate, exstipulate, petiolate, lobed; inflorescence capitulum, axillary, peduncled to sessile, heterogeneous with yellow glabrous florets; involucre two or more, seriate, ovate to elliptic-linear; inner florets 5-lobed, bisexual; stamens 5 with free epipetalous filaments; style bifid.

Cultivation

Climate

It is a delicate plant and hence cannot withstand heavy rains. It prefers light drizzles, bright sunshine, and a mild winter with no frost and heavy morning dew during the growing season. Cloudy weather and rains during flowering and seed ripening stages adversely affects the yield. However, the crop can be grown round the year for the purpose of garlands or flower pots. It is cultivated in the month of November for the purpose of oil extraction. Slight rainfall, sunlight, medium cold and morning dew is beneficial for this. Avoid higher temperature and heavy rains particularly in flowering.

Soil

The plant grows on various types of soils ranging from sandy loamy soil to black cotton soil but humus rich red loam soils are ideal.

Propagation

It is propagated by seeds. About 1.5 kg per ha seeds are required. Seeds are short-viable and hence cannot be stored for long.

Nursery Raising

Prepared 2m × 1m nursery bed and incorporated FYM @10kg/bed. Seeds are sown as dry or after wetting them along with sand for 48 hours. In the latter case, the seeds are mixed with sand @4-5 times their volume. Add water in this mixture for wetting of sand. It is tied in cloth bag and stored in warm place for 48 hours. It hastens sprouting and emergence of radicle. The nursery bed is hand watered and sprouted seeds are broadcasted all over the bed. When the seeds have settled down, a thin layer of sand is spread over just enough to cover them. Seeds germinate in about 3-4 days. If dry sowing of seeds is done then they should be mixed with sand in the ratio of 1:10 and broadcasted homogeneously. A thin layer of seeds spread to cover the seeds and the beds are hand watered twice a day. By this method seeds are germinated in 4-5 days.

Time of Transplanting

The seedlings are ready in 6-8 weeks from the date of sowing. When the seedlings are 10-12 cm tall they are transplanted to the main field at 15 × 7.5 cm spacing. After transplanting, irrigations should be provided for 10 days interval.

Fertilizer and Manures

Before transplanting, 12-15 tonnes of well decomposed FYM and 40 kg/ha each of phosphate and potash are incorporated into the soil. N is applied at 120 kg/ha in 3 equal splits for the main crop and 1 for the ratoon crop. First dose is given at10 days after transplanting and the subsequent two doses at 15 days interval thereafter.

Interculture Operations and Irrigation

Before transplanting irrigation should be done a day before. Light irrigation should provided to plants after transplanting. The crop is then irrigated weekly. Two weedings are carried out in the main crop and one in the ratoon crop.

Harvesting and yield

For oil production, the crop is harvested during last week of February or 1st week of March when a large number of flower buds start opening. Flower to plant ratio at the time of harvest is reported to be important in davana. Harvesting is done by cutting the whole plant with sickle at a height of 10 cm from the ground. The herb yield is 8-10 tonnes/ha and Oil yield is 12-14 kg /ha. After harvesting, crop is dried for 2-3 days in shade and the oil is extract through vapourization.

Plant Protection

Pests

No serious attack of insect pests has been reported in Davana. However, some insect pests like leaf-eating caterpillars, aphids and termites have been observed in the crop.

The application of Aldrex control attack of termites while the spraying of any systemic insecticide like 0.05% of Monocrotophos or 0.2% of Dimethoate will control the incidence of aphids and the spraying of contact insecticides like 0.2% Metacid will control leaf-eatingcaterpillars.

Diseases

1. **Damping off**

It is serious at the nursery stage, particularly when high humidity and cloudy weather conditions are existed. This disease, caused by *Rhizoctonia* species. It is usually severe at the nursery stage, particularly when there is high humidity and cloudy weather conditions, resulting in the heavy mortality of seedlings.

Control

i. Controlling the irrigation may reduce the incidence of the diseases.

ii. Treating the seeds with Emisan @ 0.2% or Ridona MZ at 0.1% before sowing and then drenching the seed-beds with the above chemicals a week after germination controls the disease.

iii. Adjusting of sowing time is also important for avoiding the disease.

4

Production Technology of Plantation Crops

4.1 COFFEE

Botanical Name	*Coffea* spp.
Family	Rubiaceae
Origin	Ethiopia
Chromosome number	*C. arabica* Linne – 4n=44 (tetraploid) *C. canephora* Pierre- 2n=22 (diploid) *C. liberica* Hiern - 2n=22 (diploid)
Commercial value	Mature berries/fruits
Chemical constituents	Caffeine
Research Institute	Central Coffee Research Institute (CCRI) Coffee Research Station, Balehonnur, Chickmagalur District, Karnataka.

Importance

It is one of the most valuable primary products in world trade, in many years second in value only to oil as a source of foreign exchange to producing countries. Its cultivation, processing, trading, transportation and marketing provide employment for hundreds of millions of people worldwide. Coffee is crucial to the economies and politics of many developing countries; for many of the world's Least Developed Countries, exports of coffee account for more than 50 percent of their foreign exchange earnings. Coffee is a traded commodity on major futures and commodity exchanges, most importantly in London and New York.

Caffeine content in *C. arabica* Linne –1.0-1.5%, *C. canephora* Pierre- 2.0-2.5%, and in *C. liberica* Hiern – 1.4-1.6%.

Distribution

Coffee originated in Africa, in the Ethiopian region known as *Kaffa*. From Ethiopia, coffee was taken to Arabia and finally brought into Europe in the mid 1600s first in Italy and later in England. Coffee reached America in 1607. An important milestone in 1727 was the planting of coffee in Brazil, which is now the world's dominant producer. In 1893, the coffee from Brazil was introduced into Kenya and Tanzania. The first coffee in South Africa was planted in KwaZulu-Natal at a missionary station during the late 1880s. It was introduced in India by a Muslim pilgrim, Baba Budan on the hills near Chikmangalur. Ninety two percent of Indian coffee is grown in the states of Karnataka, Kerala and Tamil Nadu, the rest in small sectors of the states of Andhra Pradesh and Odisha, as well as in some of the North Eastern states. Forty nine percent of coffee in India is arabica and fifty one percent robusta.

Description of Plant

Coffee is a short day plant. It can grow to a height of 10 to 15 m at maturity depending upon species, but is kept at 3 m in plantations for harvesting purposes. Coffee plant has two types of shoots. A prominent vertical stem with horizontal branches arising from it, in pairs opposite to each other (Fig. 1&2). The stem goes straight up and is called orthotropic (which means vertical). Multiple stems within a tree, grown from suckers, can also be found. Lateral branches spring on opposite sides of the stem and are called plagiotropic (which means horizontal). It suckers grow vertically like the main stem. 1^0 branches give rise to the 2°, 3°, 4° branches. The axillary bud provides the main cropping wood for the plant. The shrubs remain productive for 15 to 20 years.

Leaves are simple, elliptical shiny, dark green, waxy and up to 15.24 cm long. They have petiolate with persistent stipules. Their underside is marked by small cavities or domatia. Flowers are white to pinkish, fragrant and arranged in glomerules of 3 to 16, which, in turn, are grouped together in the axils of the leaves or above the leaf scars. The flowers are ephemeral, withering a few hours after they have bloomed.

Flowers occur in large bunches on old-growth wood, are generally self-fertile and will produce fruit without pollination. Flower initiation takes place between September-March. Flower buds under goes a period of clemency at time of drought condition, coupled with high temperature, long day and highlight intensity, prevailing in dry months. No vegetative growth occurs during dry months. Deformed flowers sometimes occur, called starflowers. Such flowers are sterile, this means pollination cannot take place. Therefore, no cherries will grow from them. Common reasons for the occurrence of starflowers are: too high temperatures and insufficient difference between seasons.

The fruit is a fleshy berry, in which two seeds are imbedded. Abortion of one ovule due to non fertilizing leads to a formation of a single seeded fruit called peaberry. Sometimes three or more seeds may be present due to trilocular ovaries or false polyembryary and is often called triangular seeds. Occasionally, formation of more than one ovule per locule is seen and such seeds are known aseleplant beans. The coffee fruit is oval, similar in size and shape to a small olive and turns red from green during ripening. Blossoming and fruit setting occurs mainly 2 to 3 times a year. About 6 to 7 months are required to ripen the fruit. The intervening period between the flowering and ripening of the fruits varies depending upon climate, latitude and altitude.

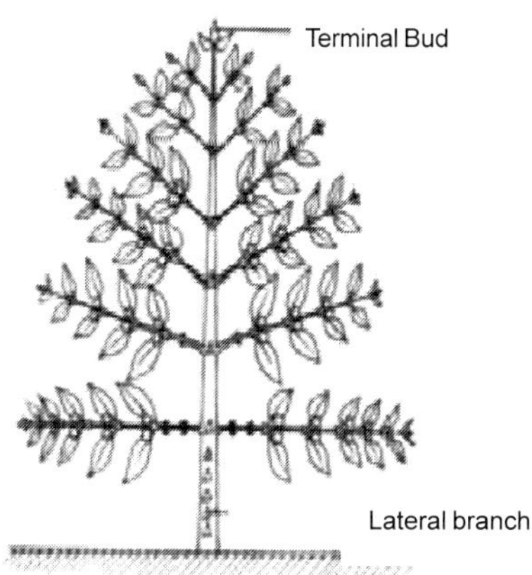

Fig. 1: Structure of Arabica Tree

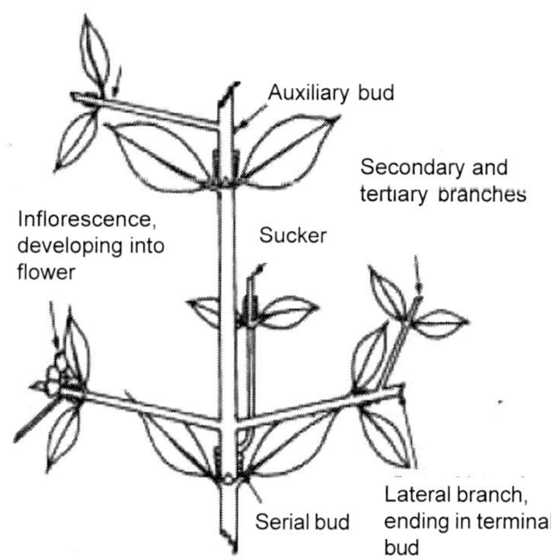

Fig. 2: Detailed Structure of Arabica Tree

Cultivation

Climate

Temperature is an important aspect in coffee production with an average temperature ranges of 12° C to 26° C. Extreme temperature variation usually affects the crop and the coffee bush. Low temperatures retard growth and below 12° C the growth of the plant is inhibited. Plants cannot tolerate frost and heavy winds. Coffee is sensitive to water shortages and adequate well-distributed precipitation of about 1500 mm a year is required.

Soil

Ideal soil type for coffee is sandy loam. Soil that are deep, well drained, slightly acidic in reaction (optimum pH is between 5.0 and 6.0) rich in organic matter content is more suitable.

Species Details

Though the genus consists of 70 species. But spp. of economic importance are as follows.

1. *C. arabica* Linne (Arabica coffee) - 65% of world production
2. *C. canephora* Pierre (Robusta coffee) - 35% of world production
3. *C. liberica* Hiern (Tree coffee)
4. *C. excelsa*

1. Arabica *(Coffea arabica)*

Coffea arabica is a valued species and has been grown and selected for several centuries. This is a glossy leafed shrub or small tree. The leaves are relatively small and the flowers fragrant and white. Arabica coffee usually receives a premium for its superior flavour and aroma but susceptible to coffee rust. It currently represents three-quarters of the world coffee production. It originates from Arabia, and thrives in land that is rich in minerals. The Arabica makes a flavoury, full-bodied coffee that is sharp in taste with a rather low caffeine content. In fact, some low-quality Arabica species are inferior to the best *Coffea robusta* varieties. Arabica beans look slightly elongated and have a range of greenish-blue shades. Arabica is more suited to higher cooler climates i.e. at an elevation of 1000-1500m above mean sea level, with an annual rainfall of 1600-2500 mm, 15-25°C and 70-80% relative humidity. Soil should be deep friable, porous rich in organic matter, moisture retentive, slightly acids pH 6-6.5. It needs medium to light shade depends on elevation aspect and a gentle to moderate slope is ideal.

2. Robusta *(Coffea canephora)*

Coffea robusta leaves are bigger than Arabica and the tree can remain productive for a longer time. Cherries and beans of Robusta are easily distinguished from Arabica because their shape is more rounded. It is a variety that can grow to more than 12 m in height. It grows quickly in altitudes up to 600 m, and is more resistant against coffee leaf rust (*Hemileia vastatrix*), root knot nematode (*Meloidogyne exigua, M. incognita, M. paranaensis,*

Pratylenchus brachyurus and P. coffea) and the coffee berry disease *(Colletotrichum kahawae).*This variety was discovered in the Congo in 1898 and is widely spread, especially in Africa, Asia and Indonesia where the climate is unsuitable for the cultivation of *Coffee arabica.* It represents approximately 25% of the total world coffee production. Because of their higher caffeine content (about twice as much as Arabica) and strong character, Robust as are used mostly as blends. Robusta beans are typically small, rounded and brownish yellow in appearance. In general, they can thrive in hotter lowland areas *i.e.* at an elevation of 500-1000m above sea level with an annual rainfall of 1000-2000 mm, 20°-30° C temperature and 80-90% relative humidity. It need uniform thin shade. Robusta coffee is preferred for instant coffee production due to higher soluble solid extraction.

3. Liberica *(Coffea liberica)*

This is a larger tree with large leaves and berries. The leaves of Liberica are bigger and more rounded than Arabica and Robusta. Liberica trees can reach heights of over 10m. Their yields are typically low. The coffee brew of Liberica beans is very bitter and sour. The inconvenience of harvesting tall trees and the low quantity and quality of the crop has reduced the amount of Liberica. It can tolerate hot and wet conditions. The coffee produced is bitter. It is having rust resistance. This is preferred in Malaysia and West Asia.

4. *C. excelsa*

It comes from the more continental and drier parts of Central Africa, mainly the Central African Republic. Genetically, *Coffea liberica and C. excels* are now considered as a single complex.

Propagation

1. Seed propagation

The propagation of seed from the Arabica varieties gives plants that are true to type. Arabica trees are self-pollinating.

Healthy and mature fruits of normal size and appearance 3/4 to full ripe are harvested for seed purpose. Cherries that are ripe during the peak of the harvest are usually the best for propagation purposes. When harvesting the cherries for propagation only pick the big, healthy red cherries, thereby increasing the chance of successful germination and propagation. Cherries from the middle part of the tree are preferable over those from the bottom and top part of the tree as these developed in a lightly shaded environment leading to better quality.

Harvested cherries are put in water, green berries and the floating ones are skimmed from the surface. Processing can be done by hand. The ripe cherries are squeezed by hand until the beans pop out. Then sort out the beans. Pulping does not remove all the pulp. Final cleaning of the parchment is done by fermentation. After pulping pack the beans in plastic bags close them, and leave them overnight (approximately 10 hours). After fermentation the beans should be washed 2 -3 times with clean water to remove all remaining pulp.

The beans (or parchment coffee as it is called after fermentation and washing) are then dried in sun. The beans spread the parchment in a layer of 2cm deep on a clean surface and turn the beans every second hour on the first day. At night and during rain the parchment should be moved inside to avoid rewetting. At around 25% moisture content the drying is finished. The parchment now looks white and is tight around the beans. The colour of the beans is bluish. The final step of processing is the removal of pea berries, broken and insect invested parchment as well as; parchment with more than 2 beans (Kuit, 2004).

2. Vegetative propagation

Vegetative propagation is done by cuttings and grafting. To obtain a good plantation, coffee trees within a plantation with high yield and resistance to leaf rust characteristics can be selected and propagated vegetatively by means of leaf cuttings. After cutting, they should be put in water immediately and transferred as soon as possible to a nursery bed. The best time to do this is when the weather is cool. Generally, root systems developed from cuttings are less vigorous or well-developed as trees propagated by seed. Therefore, this technique should not be used in areas where drought is likely to occur. A coffee plantation can, by this method, gradually be replaced with better trees or a new plantation consisting of potentially good trees established. The purpose of grafting is to combine several traits of different trees into one. For example, some trees have very good root systems, but have low yields. A potential advantage of vegetative propagation is that the new trees will have the same genetic characteristics as the mother tree.

Nursery preparation

Ideally the nursery bed should be of size 6m × 1m × 15 cm. 4 baskets of cattle manure, 2kg finely sieve lime, 400 g of Rock phosphate is added to beds.

Before sowing, the parchment should be submerged in water of 50-60°C (at this temperature the water is just bearable to touch) with 2% of lime (add 20g of lime to 1liter water), stir it well and leave it for 24h. Lime will increase the pH level of the water, stimulating germination. The parchment should be sown at a depth of 1cm (Kuit, 2004). About 1.5 to 2.5cm is commonly recommended distance between seeds. A thin layer of fine soil and paddy straw is used to cover the seeds to prevent the damage from sunlight. Seed germinate with in 45 days. Seedlings are transplanted into second nurseries or poly bags. To ensure good root development the bags should be sufficiently large. Coffee seedlings are transplanted to 23 × 5 cm^2 with 150 gauges polythene in February-March when they are button of topped stage. Shorter bags can inhibit proper root development, and more narrow bags will result in a too close spacing of the seedlings, which then will become long and thin. These are more susceptible to wind damage after transplanting the seedlings to the field. Each bag should have some holes in the bottom to allow for drainage. At the time of transplanting taproot should be nipped off slightly. Seedlings are in secondary nursery. Regular watering, manure in two months is necessary.

Transplanting

The seedlings are gently removed from the nursery. Seedlings with double taproots can be maintained, but with one root should be removed before planting. Other than that only seedlings with a straight taproot should be used.

In 10-12 weeks after sowing, the terminal bud will develop into a pair of primary leaves. Every new pair of primary leaves takes 3-4 weeks to emerge. The best time for transplanting is during June or September when seedlings are 16-18 months old having 5-7 of pairs of leaves and an overall height of 20-30 cm.

Land preparation

Best time for land preparation is before the onset of monsoon. Wild shade trees are essential. Land divided into small blocks. Each planting hole should be 0.5 × 0.5 × 0.5m. Straight row planting with an east-west orientation is the recommended layout. The top 20cm of the soil should be kept separate during the digging. This can later be mixed with fertilizers and manure or compost before it is returned to the hole. Hole is made in the pit and seedling should be inserted into it and hole is filled. Filling is done at 3cm above from of ground level to prevent water stagnation. The spacing is kept 2.0-2.5m for Arabica and 2.5-4.0 m for Robusta. The planting holes of the young trees have been covered with old weeds or other material such as coffee pulp, husk or straw. This will lessen the impact of rain on the soil, and maintain better soil moisture content.

On sloping land with steepness exceeding 5% contour planting should be practiced. This is to prevent excessive loss of fertility and nutrients by erosion.

Training and pruning

Training and pruning are necessary for strong frame work and it promotes bearing of wood.

Training

Single stem system: When plant reach at 75 cm in Arabica or 110cm in Robusta it is topped. It is done to restrict vertical growth, facilitate lateral spreading to increase bearing area.

Pruning

It is done just after harvesting till the onset of monsoon to divert the vigour of plants to certain parts by pruning other parts. Pruning involves:

i. Centring- Removal of vegetative growth up to 15cm radius from the centre up to the first node of all primary branched.

ii. Desuckering- Removal of orthotropic branches arising from the main stem.

iii. Handling- Removal of small sprouts.

iv. Nipping- Growing tip of 10 branches is removed to encourage secondary and tertiaries.

Soil Management

Digging: In the new clearing, the field is given a thorough digging to a depth of 35-45 cm towards the end of monsoon. All weeds and vegetative debit get turned and buried in soil.

Scuffing or Soil stirring: In established coffee fields stirring is done in beginning of dry period, control weed and conserve soil moisture.

Trenching: It is done between coffee rows. Trenches are 50 cm broad and 20cm deep.

Mulching: For optimum soil moisture, erosion control and increase fertility, mulching is followed.

Weed Control

Weeds should be controlled before planting either mechanically or with a herbicide. Hand weeding could be done 3-4 times in a year. During the early stages and to facilitate establishment, mulch could be placed around the young seedling for weed control. Mulch should not be placed closer than 100 mm from the main stem.

Irrigation

The main purpose of irrigation is to supplement rainfall. Keep the soil moist, but not too wet. Over-irrigation will enhance the risk of fungal diseases. Irrigate the plants twice a week in hot weather, and once a week in cool weather.

Shade and its Management

It is a pseophyte plant. Lower/temporary canopy prefer *Erythrina lithosperma* Planted along with coffee. Silver Oak (*Grevillera robusta*) used as a temporary shade. Permanent shade like Ficus, *F. retusa, Dalbergia latifolia, F. glomerata* are used. Shade trees are generally planted about 12-14m apart. It is advisable to plant a large number initially and out as the trees grow and spread out. Most convenient time of regulate shading is after pruning and training

Fertilizer and Manure

Coffee plants produce every year. Fresh wood for the succeeding crop, hence they require a regular supply of nutrients. Due to leaching of nutrients, it requires extra nutrient application. The peak periods of demands for nutrients are at the flowering time, fruit set and maturation of the crop. These are applied in a broad circular band about 30cm away from the stem. Foliar spray with urea, is also followed. Bordeaux mixture is sprayed at flowering time and fruit setting. Fertiliser application should be done four to eight weeks after planting and every year in August/September for plant establishment. Nitrogen is important to increase growth, prevent leaf fall thereby maintaining the photosynthetic area of the plant. Its shortage may affect production. Potassium, zinc and boron are some of the essential elements in coffee production. Boron and zinc deficiencies result in shortening of internodes and smaller leaves. Potassium deficient plants defoliate. Potassium and nitrogen should be applied from September to April. Zinc should be applied when necessary during November and February annually. Boron is needed for cell division and it can be applied at the beginning and towards the end of the flowering period.

Harvesting

Fruits should be picked as and when become ripe to get better quality. Arabica comes for harvesting earlier since they take 8-9 months for fruit development while Robusta take 10-11 months. Picking is done by hands.

First picking- First picking consist of selective picking of ripe berries often seen in the outer portion of the nodes and is called fly picking. Thereafter, there will be 4-6 pickings at 10-15 days interval and final harvesting i.e. stripping consist of picking of soil refraining green berries on the plants.

Processing

There are two ways by which coffee can be processed - dry ('natural') processing and wet ('fermented and washed') processing. In most cases, wet processing is regarded as producing a higher quality product. However, some areas prefer dry processed coffee for its 'fuller' flavour.

1. Dry processing

This is the simpler of the two methods and is popular in Brazil to process Robusta coffee and in Sri Lanka to process Arabica coffee. The coffee cherries are dried immediately after harvest. They are usually sun dried on a clean dry floor or on mats. The bed depth should be less than 40mm and the cherries should be raked frequently to prevent fermentation or discoloration. However, there are problems associated with this method. The most serious problem is dust and dirt blown onto the produce. Hence, nowadays Solar and Artificial driers are used for this purpose.

2. Wet processing

In this method the cherry is squeezed in a pulping machine or pestle and mortar which removes the outer fleshy material (mesocarp and exocarp) leaving a bean covered in mucilage. This mucilage is fermented and dispersed. The bean is washed and dried.

Plant Protection

Diseases

1. Coffee leaf rust

Two species of *Hemileia* fungus have been reported as causing rust on coffee, *viz. H. vastatrix* and *H. coffeicola*, but *H. vastatrix* is regarded as the more potent pathogen of the two.

Symptoms are mainly confined to the leaves and start as small, yellow spots on the under surface of the leaves. As the lesions increase in size, they darken to a typical orange colour. Brownish blotches eventually make their appearance on the upper surface of the leaves. As the lesions mature, the centre of the lesions on the lower surface becomes necrotic. Heavy infections can lead to the defoliation of the trees. With *H. coffeicola*, the orange powdery lesions which cover the lower surface remain green for some time.

Control

a) The use of available contact and systematic fungicides for the control of this disease is recommended. Spray Bordeaux mixture 5%.

b) Growing resistant varieties.

2. Black Rot or Coffee berry disease

The fungus *Glomerella cingulata* causes a dark-brown rot, which destroys the beans and causes the berries to dry out. The spots appear on the lateral surface of the green berry, which enlarges rapidly to produce roughly circular, slightly sunken, dark-brown spots. Several spots develop on the fruit to form irregular, necrotic areas. Infection can also start at the base of the berry. Pinkish masses of asexual spores (Collectotrichum) can be produced on the surface of the lesion. Diseased berries, with or without their fruit stalks, often drop. Lesions, which tend to become ash grey (except for a dark brown margin) tend to develop at the onset of the dry season and are referred to as scab lesions.

Control

a) Spray 1% Bordcaux mixture

Varieties

Kents	It is the earliest variety of Arabica, selected by an English planter of the same name during the 1920s. This variety remained popular with the planting community till the 1940s, because it was less susceptible to rust. Today, it is grown in a few areas but it is still known for its exceptional cup quality.
S.795	It is the most popular Arabica selection released during the 1940s with high yields, bold beans, superior quality and relative tolerance to leaf rust. This selection was developed using 'Kents' Arabica, known for its high quality. Even today, the S.795 is a favourite with the planters and is a widely cultivated Arabica variety. It has a balanced cup with subtle flavour notes of Mocca.
Cauvery	Popularly known as Catimor, Cauvery is a descendant of a cross between 'Caturra' and 'Hybrido-de-Timor'. It is a natural mutant of the famous Bourbon variety. Thus, Cauvery inherited the high yielding and superior quality attributes of Caturra and the resistance of 'Hybrido-de-Timor'.

Selection 9	It is a derivative of a cross between an Ethiopian Arabica collection, 'Tafarikela', and 'Hybrido-de-Timor'. It has inherited all the superior cup quality traits of Tafarikela.

References

Kuit,M. (2004) Manual for Arabica cultivation. Coffee Handbook. P P P Project on "Improvement of Coffee Quality and Sustainability of Coffee Production in Vietnam".pp18-20.

4.2 TEA

Botanical Name	*Camellia sinensis* L.O. Kuntze. (China Jats) *Camellia assamica* (Assam Jats)
Family	Theaceae/Camelliaceae
Origin	Highlands of South West China, Myanmar and North east India.
Chromosome number	2n =30
Commercial value	Leaves

Importance

Tea is one of the most popular and lowest cost beverages in the world and consumed by a large number of people. Owing to its increasing demand, tea is considered to be one of the major components of world beverage market. The tea industry in India is more than 170 years old, since the sending of 12 chests of tea, commercially cultivated and manufactured in Assam, for the London auction in 1838. Tea is traded as:

a. Black tea: It is a fully fermented tea.

b. Green tea (East Asia, Arabian countries): Through heat treatment (in pans or with steam), polyphenol oxidase (enzymes) in the fresh leaves are inactivated. Only then is the product rolled and dried (often frequently). Fermentation is suppressed by deactivating these enzymes, and the leaves retain their olive green colour.

c. Oolong tea (China, Taiwan): The fermentation process is halted at an earlier stage and it is partly fermented tea.

d. Instant teas: Instant teas are made either from low-quality teas (fermented and dried), or from non-dried tea in a special process directly after fermentation. Instant teas lose much of their aroma during the extraction (only hot water extraction is permitted) and subsequent freeze-drying processes.

Distribution

Tea cultivation is confined only to certain specific regions of the world due to specific requirements of climate and soil. Majority of the tea producing countries are located in the continent of Asia where China, India, Sri Lanka are the major producers. African tea growing countries are located mostly around the tropical regions where Kenya, Malawi, Rwanda, Tanzania, Uganda are major producers. Apart from these regions, some quantities of tea are also being produced in South America (Argentina, Brazil and others), the Near East (Iran and Turkey) and Russia. In India, the major tea growing states are Assam, West Bengal, Tamil Nadu and Kerala. Other states such as Bihar, Himachal Pradesh, Karnataka, Orissa, Tripura and Uttaranchal also grow tea in small areas. Tea plantations in India are traditionally large holdings but in recent years small scale cultivation has gained prominence and at present there are more than 200,000 small tea growers who account for 24 percent of the production. India is the largest producer and consumer of black tea in the world.

Botany

The cultivated varieties are hybrids of the original tea plants *Thea sinensis and Thea assamica. B*otanically, the two extreme varieties are distinguished *viz., Camellia (Thea) sinensis* var. *sinensis* and *Camellia sinensis* var. *assamica.* The variety *sinensis* also called China tea, is suitable for marginal areas of the tropics especially suited to hilly regions. It is resistant to drought, and can tolerate short periods of frost (yet only has a low tolerance of shade). It is more drought tolerant and can survive short frost periods. The leaves are small (upto 9 cm long), tough, leathern with edges not pointing with a strong flavor. Contrastingly, the var. *assamica* also called Assam tea, is a purely tropical crop, and reacts sensitively to drought and the cold (yet only has a high tolerance of shade). The leaves are big (upto 35cm long), soft,without edges, pointing with less flavour and high yielding. It grows as an evergreen tree to a height of 8-15m with strong taproot under natural conditions.

Description of plant

Assam Tea	China Tea
It is a tree	It is a shrub
Light to medium green	Dark green colour
High yield and medium quality	Low yield but good quality
Susceptible to drought and frost	Hardy and resistant
Sparse flowering	Profuse flowering

Soil

The soil should be deep, well-drained and aerated. Nutrient-rich and slightly acidic soils are best with optimum pH-value 4.5-5.5.

Climate

Tea requires an average annual temperatures of 18-20°C, an average daily amount of sunshine of 4 hours per day, as well as a minimum of 1600 mm of rainfall distributed evenly throughout the year. The optimum temperature for growth and development ranges from 20 - 27°C. Relative humidity should lie between 70% and 90%. Its performance is excellent at elevations ranging from 1000 - 2500 m. Tea plantations in windy regions should also be protected by wind breakers e.g. hedges, to reduce the intensity of evapo-transpiration

Cultivation

Land Preparation

The land is cleared of the roots of the fallen trees and drains are taken at suitable intervals depending upon the slope to conserve the soil. In the olden days, up and down system of planting at 1.2x1.2 m are followed. Presently, contour planting of tea either in a single hedge or double hedge system is followed.

Propagation

Tea can be propagated by seeds or vegetatively by cuttings. Only seedling trees develops an extensive taproot but they are less yielding, hardy and does not give a uniform stand because of the heterogeneous source material. The vegetative propagation offers a large choice of improved clonal tea varieties. Plants have high yield potential, they are more sensitive and demanding than seedling tea.

Nursery

Seeds and cuttings are raised in a nursery under controlled circumstances over a period of 1-3 years. The nursery soil should be well drained and deep loam in nature with pH of 4.5 to 4.8 and free from soil borne pathogens and nematodes. The soil and sand used in the preparation of rooting medium. To reduce the soil pH nursery beds are treated with Aluminium sulphate. Nursery beds of one metre width and about 8 cm height of convenient length depending upon the land available. Then the beds are drenched with 2% solution of Aluminium sulphate applied at 10 litres/2.5 m^2 of area. Over this another layer of soil of 8 cm height is spread and again drenched with equal quantity of water twice. Then the soil is allowed to dry and the pH is checked before use in the nursery.

Preparation of cuttings: Cuttings are taken on April - May and August - September. Semi hard-wood cuttings are prepared with one leaf and an internode with a slanting cut at the bottom. Healthy and vigorously growing high yielding bushes should be selected. Apply the following mixture before taking the cuttings.

- 0.5 % $AlSO_4$ + 1 % $MgSO_4$ (before 3 weeks)
- 2 % Zn SO_4 (before 2 weeks)
- 1 % Urea (before 1 week)

Planting of cuttings

The sleeves are watered thoroughly and holes are made in the soil. The cuttings are inserted in the hole and the soil around is pressed firmly to avoid airspace followed by watering. These cuttings take 10 - 12 weeks for rooting.

Season and planting

The main season for planting of tea is May - June or September – October.

Methods of planting

Single Hedge System: In this method, the spacing adopted is 1.20 × 0.75 m accommodating 10,800 plants/ha.

Double Hedge System: In this method, the spacing adopted is 1.35 × 0.75 × 0.75 m accommodating 13,200 plants/ha.

Manuring and Fertilization

A high amount of nutrients are lost through the continual plucking of tea leaves. Application of fertilizers should be done before the onset of monsoon. Fertilizers should be broadcast around the drip circle avoiding contact with the collar. Fertilizer and manure schedule is given below:

Table: Manures and Fertilizers schedule

Year of application	N kg/ha/year	K kg/ha/year
I year	180	270
II year	240	360
III year	300	450
IV year onwards	300	300

Intercultural Operations

Pruning the tea bushes

The Pruning is essentially the artificial removal of the leaf bearing branches of the plant. The operation is aimed at keeping the size and vegetative vigour of the plant in a condition most conducive for maximum vegetative growth.

Regular pruning of the tea bushes is one of the most important measures in cultivating tea. Pruning intervals depends on the site and plucking system. Usually, the bushes are pruned back to a comfortable plucking height every three years, and then radically cut back every 15-20 years (to a plant height of 30-40 cm) for rejuvenation and frame sanitation is practised, followed by a period of recovery. Collar pruning, reaching down to the soil, is utilised to rejuvenate the tea plants. No fundamentally different pruning measures are used to those carried out on conventional tea plantations. Shorter pruning intervals with less pruned off will facilitate the decomposition of pruning material by the soil life.

Weed Management

Mulching methods can be especially recommended to effectively combat weeds (and erosion-prevention). Hoeing is not recommended on those sites at risk from erosion. Motor scythes could be used for weed control.

Shading tree management

The shading trees need to be continually thinned out to create and maintain an optimum amount of shade. Thinning out will also help prevent infestations of blister blight (*Exobasidium vexans*), which thrive under too shady and moist conditions. The shading trees should be trimmed to prevent blister blight developing directly before the rainy season (monsoon).

Harvesting

Plucking

The process of harvesting tea by selective hand picking of tender shoots (upto 30 kg per day per plucker) is known as plucking. The tender shoots harvested consisted of the terminal bud, the internodes and 1, 2 or 3 leaves immediately below the bud. Shoots are plucked throughout the year at regular intervals (4-10 days). It is the most expensive and labour intensive field operation of tea cultivation. Use of machineries is limited in this operation because of the natural circumstances like steep slopes in which tea is grown.

Yield

Processing

The plucked leaves are stored in loose manner and to avoid leaf damage as well as fermentation are brought immediately to the factory. The main steps in tea processing are Withering, Rolling, Fermentation, Drying and Sorting. Depending on the maceration process (referred as orthodox) CTC (Crushing Tearing Curling) and LTP (Lawric Tea Processor) different tea qualities are obtained.

Minimum content levels

Dry matter	min.	93g / 100g tea
Extract contents	min.min.	32g water soluble constituents / 100g dry matter 26g water soluble constituents / 100g (for tea of Turkish or Russian origin)
Caffeine content	min.	1,5g / 100g dry matter
Total ash content	min.max.	4,0g / 100g dry matter8,0g / 100g dry matter
Ash insoluble in hydrochloric acid	min.	1,0g / 100g dry matter
Water soluble ash	min.	45,0g / 100g dry matter
Raw fibres	min.	16,5g / 100g dry matter

Source: Tea-*In* : Organic Farming in the Tropics and Subtropics Exemplary Description of 20 Crops. Naturland e.V. – 1st edition 2000 pp 12.

Insect-pests

Pest

Red coffee borer

Symptoms: Larva bores the young stem, tunnels downward.

Control

Affected stems are cut upto the healthy and heptachlor ispoured in the hole using ink filler and pluged with clay paste.

Shot-hole borer

Symptoms: Grubs make a typical shot-hole on the branches and inside gallaries. A serious problem in low and mid elevation areas.

Control

Badly affected branches are pruned off. Heptachlor 20EC is sprayed @ 8.51 in 675 lit. of water /ha on the pruned frame and prunings.

Red spider mite

Symptoms: Infests upper surface of mature leaves.

Control

Apply tetradifon (8EC) 1-1.25 l/ha.

Thrips

Symptoms: Leaf surface becomes uneven, curly and metty, exhibiting parallel lines of feeding marks on either side of the midrib.

Control

Endosulfon 1 lit/ha.

Nematodes

Symptoms: Occur in tea industry, infested roots develop galls.

Pre heat treatment of soil media upto 60-80 °C and application of carbofuran 3 G @ 80 g/ cubic metre of medium.

Diseases

Blister blight (*Exobasidium vexans*)

Symptoms: Infected tender leaves and stem and develops translucent spot. Cloudy and wet weather favour infection.

Control

Copper oxychloride 350g. in 67 lit. of water with power sprayer for pruned field at 3-4 days intervals. In the plucking field 210 g copper oxychloride+210g Nickle chloride in 45 lit. of water/ ha at 7 days interval.

Black root diseases (*Rosellinia arcuata*)

Infested roots show black mycelium on the roots, white star shaped mycelium between bark and wood and black lead shot like perithecia seen on collar region.

Control

The soil may be drenched with Dithane M-45 @30g/10lit.

Varieties

The tea varieties that are cultivated are all hybrids of the original tea plant *Thea sinensis* and *Thea assamica.*

Pandian, Sundaram, Golconda, Jayaram, Evergreen, Athrey, Brookeland, BSS 1, BSS 2, BSS 3, BSS 4, BSS 5

4.3 COCOA

Botanical Name	*Theobroma cacao*
Family	Sterculiaceae
Origin	Central Amazon basin (Tropical America)
Chromosome number	2n= 20
Commercial value	Seeds
Research Institute	Central Plantation Crop Research Institute, Kerala Directorate of Cashewnut and Cocoa Development

Importance

Cocoa originated from the Amazon forests as well as the rainforests of Central America (the white seed subspecies Criollo). The crop was already known to the Aztecs, who relished it as a "drink of the Gods". In 1847, an important discovery was made by John Fry in England. By adding cocoa butter to a mixture of liquor and sugar, chocolate was created, one of the confectionery industrys greatest discoveries. This is not only an easily handled product, but it is solid at room temperature and melts just below body temperature. Thus, it is a product that, when eaten, releases its flavours in an optimal manner (Anonymous, 2009). The cocoa seeds (fermented and dried beans of their pods are utilized) are highly rich in fat content, and therefore provide an energy-rich and delicious foodstuff. The seeds of the cocoa plant are mainly used to manufacture chocolate, cocoa based drinks and confectionary food items. Fresh cocoa fruit juice from the seed pulp, which is produced during the process, can be fermented into alcoholic beverages and vinegar, or made into fruit jellies. On traditional cocoa plantations in Mexico (Tabasco) the ground and dried cocoa is mixed together with maize flour, then stirred together with water, to provide a nourishing drink during the work.

Distribution

The successful cultivation of cocoa requires a special climate that is mostly found within the area bounded by the Tropics of Cancer and Capricorn. The majority of the world's crop is now grown within 10° North and South of the equator. The largest cocoa producing countries are Ivory Coast, Ghana and Indonesia. Today, Africa is the main overall cocoa supplier, with 75% of the world's cocoa crop however, United Kingdom and France are major consumer. In India it is being cultivated in Kerala, Karnataka and Tamil Nadu. The cultivation of cocoa on a large scale started in India in the early 1970s. It is mainly grown as a mixed crop in coconut and arecanut plantations and as under-storey crop in partially cleared forests. Kerala accounts for 71 per cent of the area and 80 per cent of the production. Though Andhra Pradesh has the second highest area, Karnataka ranks second in production.

Description of plant

The trees attain a height of 8-10 m (Fig. 3). The blossoms appear in the many year old wood of the leaf axil, on the trunk and on the branches (Kauliflorie). Out of the several thousand flowers on a mature tree, only a small number matures into fruits or pods. The fruits grow directly from the trunk of the tree and the thicker branches. These take some 5-6 months to grow from a fertilized flower that are pollinated by insects, mainly of the species *Forcipomyia* and *Lasioshelea.* The cocoa fruit has a cucumber like shape, and is about 25 cm long, 8-10 cm thick and weighs 300-400 g. The shell, which can be up to 20 mm thick, surrounds the sugar-rich, bitter-sweet, acrid pulp. The fruit contains 25-50 almond-shaped, bitter tasting seeds, in 5-8 long rows: the cocoa beans. After fermentation and drying, one pod produces some 40 g of beans, one bean typically weighing around 1 gram. Blossoms can appear throughout the whole year, providing no extreme drought periods or seasonal temperature fluctuations occur.

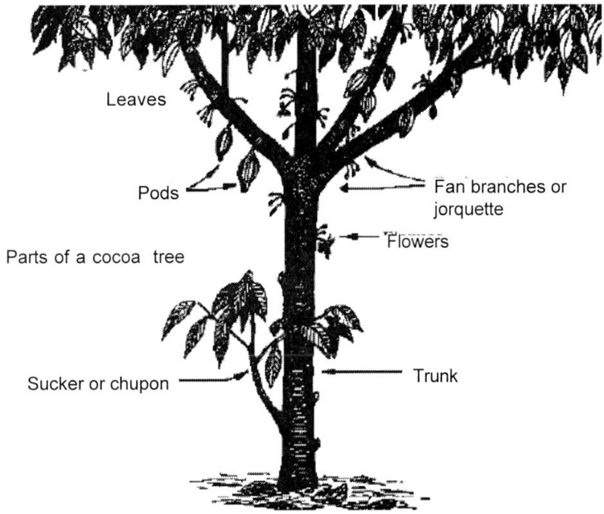

Fig 3. Parts of Cocoa Tree

Climate

Cocoa is grown from sea level up to an elevation of about 500m. It comes up best up to 300m above mean sea level. It required 15-32ºC temperature. Optimum temperature is 25°C. It cannot be grown in areas where the minimum temperature falls below 10°C and the annual average temperature is less than 21°C. The plantations can survive on 1250 mm per year with 90-100 mm rainfall per month. Cocoa needs high humidity throughout the year for optimum growth. Short drought periods can be compensated for by heavy clouds and high humidity. Dry months should not exceed 3-4 months. In regions with extensive wet periods or large seasonal temperature fluctuations, the harvesting periods are reduced to only a few months per year. In the regions with a balanced climate, and only slight temperature and rainfall fluctuations, cocoa produces fruit practically throughout the year.

Soil

Cocoa is a forest crop and so it is suited to forest soils. Hence, it is exceptionally demanding in its soil requirements. Soil should be deep, well-drained, and have sufficient water-retaining capacity. The pH-value should lie between 4.0 and 7.5, with sufficient organic material.

Propagation

Seed is the cheapest and most convenient planting material in cocoa. Seed pods may be collected from trees yielding 80 or more pods per year with pod weight 350-400g. Fresh beans from such pods should be used for sowing, as cocoa seeds should loss their viability soon after they are taken out of pods. Cocoa seeds have no respite (interval) in germination. If the fruits are harvested when overripe, then germination can already begun in the shell. A large number of already germinated cocoa seeds will not pass a quality control, for this reason, overripe and diseased fruits should not be mixed together with healthy cocoa fruits, but are to be processed separately. Before sowing the seeds are rubbed with dry sands or wood ash to remove mucilage. Beans are planted with their pointed end upwards, either in plastic bags (25 × 15cm, 150 gauge thickness) or in raised bed. If sown in raised beds, young seedlings should be transplanted into polythene bags after about 2 weeks of germination. The seedlings are ready for transplanting in field in 3-4 months or they attain a height of 30 cm.

Cocoa can also be propagated by vegetative method *viz.*, T- budding, Inverted T, forkert method of budding and micro- budding. Cocoa being a self-incompatible crop in general, the population will have lot of variations. Therefore, the need for asexual or vegetative propagation arises when the seedling progeny of selected trees are variable. Softwood grafting is found successful and suitable in cocoa to breed true to type planting materials. The method consists of cleft grafting of scions to seedlings of 75-90 days old raised in polybags. The grafts are planted in the pits as in the case of seedlings after three months of hardening. About 80 to 90 percent success is expected if grafting is done during August to October.

Establishing plantations

It requires partial shade (50% sunlight). Site selection is very important for plantation. It can be grown in arecanut and coconut gardens as a mixed crop. It can also be planted in forest lands by thinning and regulating the shade suitably. It is planted at a distance of 2.5-3m both between and within rows. It can be planted either in the beginning of the Southwest monsoon (May-June) in low rainfall areas or at the end of the monsoon (September) in high rainfall areas.

According to Ananda (2006) cocoa seedlings are to be planted in pits of 60 cm^3 filled with compost. Cocoa is planted in 2.7m × 5.4m spacing in areca garden planted with a spacing of 2.7m × 2.7m (Fig. 4). While establishing new garden a spacing of 3.3m x 3.3m can be followed for both areca and cocoa. Both the crops require shade during summer season after planting. When cocoa is to be raised as a mixed crop with coconut (Fig. 5) single hedge or double hedge system (Fig. 6) of planting can be adopted. In single hedge system, cocoa can be planted 2.7 m apart in a single row between two rows of coconut, while in double hedge system it can be planted 2.5m apart in paired rows between two rows of coconut palms.

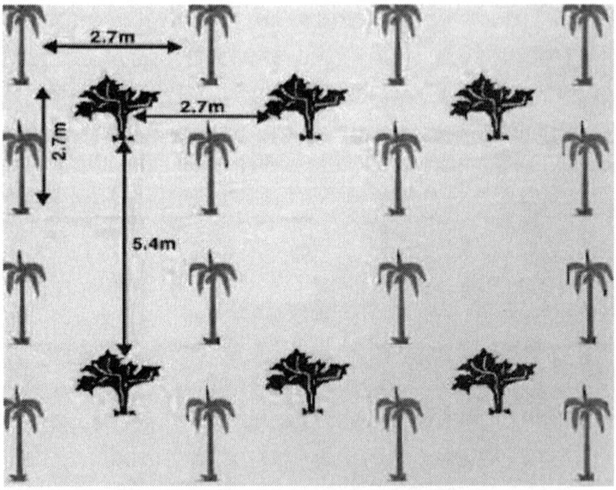

Fig. 4. Cocoa under arecanut

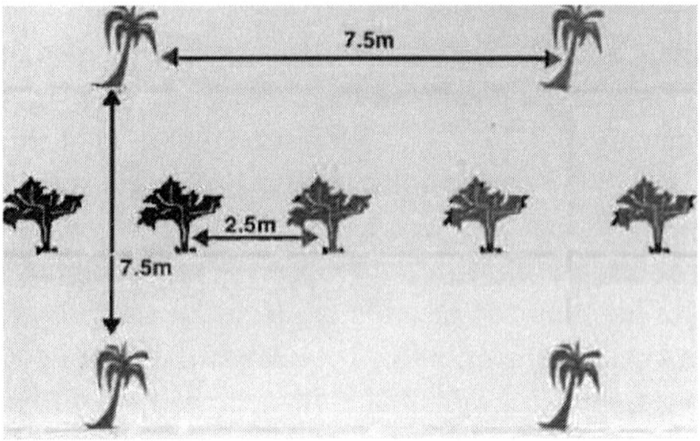

Fig. 5. Cocoa under Coconut (Single hedge)

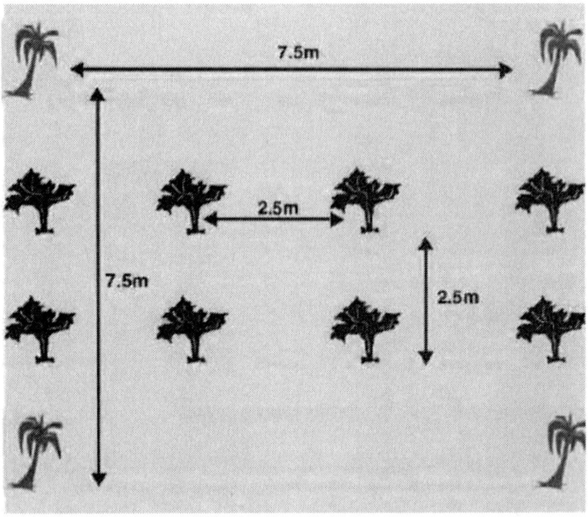

Fig. 6. Cocoa under coconut (Double hedge)

According to Asare and David (2011) in between the plantain and cocoa rows, plant cassava cuttings at a dimension of 2 × 2 m from each other and 1.5 m in between rows of the plantain and cocoa (Fig. 7).

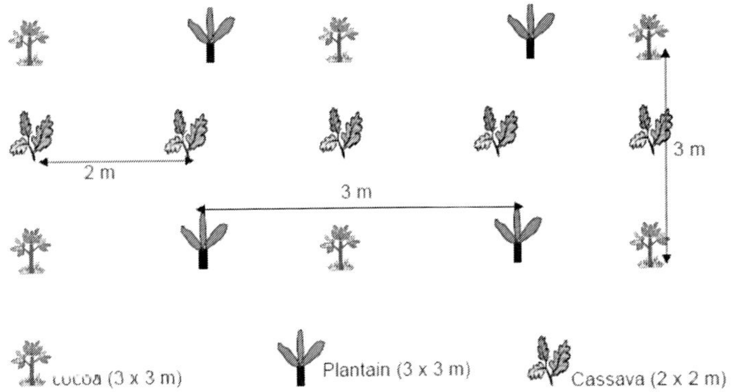

Fig. 7. Cocoa with Cassava and Plantains

Manure and Fertilizers

Annual application of 100g N, 40g P_2O_5 and 140g K_2O per tree in two equal splits is recommended. Nitrogen is given per year per plant in 2 split doses. During 1st year 1/3rd of nitrogen is given in March-April and in 2nd year 2/3rd of this dose is applied during September - October. Phosphorus and Potash are given as basal dose.

Irrigation

This plant requires continuous supply of moisture. In summer irrigation should be given weekly.

Pruning

Pruning should be done regularly to develop good shape. When the first jorquete develop at a height of about 1.5m., the canopy will form at a height convenient for harvesting and other operations. Hence, all the fans branches arising from the main stem are nipped off up to a height of about 1-15 m or cutting initial year of growth. A second jorquete may be allowed to develop, if the first one formed very close to the ground. When more fan branches develop, 1 or 2 weaker may be removed. Diseased branches should also be removed.

Maintenance

During the first three years, the cultivation measures consist almost entirely of selective regulation of the weed growth. Trees which do not loose their leaves, need to be radically trimmed during the blossoming period of the cocoa (6 months before the main harvest begins) in order to increase the amount of light. The resulting organic material should then be chopped and spread out over the soil. Mulching is important in young cocoa plantation (up to 3 years) but is not necessary in mature farms where fallen leaves act as mulch. Diseased plant parts should also be removed.

Harvesting

Cocoa harvesting starts from a second year of planting (4-5 months).Pods take about 140-160 days to mature and ripen. Each pod will have 25-45 beans embedded in white pulp (mucilage).Cocoa gives 2 main crops in a year September–January and April–June. Pods are harvested by cutting the stalk with the help of a knife.

Plant Protection

Cherelle wilt is a physiological disorder of cocoa. Cocoa produces a large number of flowers but only small percentages of flowers are successfully pollinated and few many fruits are set for the tree to carry through maturity. The young cocoa fruits, till it attains a length of 10cm, is called cherelle and over 80% of the cherelle formed on a mature tree usually wilt. The cherelle initially stop growing and a week later turn yellow and then blacken and shrivel.

Varieties

Character	Criollo	Forestero
Cotyledons	Plumpy, white when fresh and turn coloured on fermentation	Flat purple when fresh and dark chocolate brown on fermentation.
Pod colour	Dark Red	Yellow
Other pod Characters	Rough surface, ridges Prominent, Pronounced point and thin walled Band Flavour.	Smooth, icons picuous Harsh flavor, bitter taste.
Aroma duration	3 days	6 days
	Poor yield and less yield potential and hence discourages for commercial cultivation.	Good adaptability and high yield and hence recommend for commercial cultivation.

The hybrid between these two *viz* - Criolla × Forestero is Trinitario. Forestero has good adaptability in India and high yielding.

References

Anonymous (2009). De Zaan Cocoa and Chocolate Manual 40th Anniversary Edition. ADM Cocoa International, Switzerland. pp 9-11.

Asare, R. and David. S. (2011). Good Agricultural Practices for Sustainable Cocoa production: a Guide for Farmer Training. Manual no. 1: Planting, replanting and tree diversification in cocoa systems, Sustainable tree crops programme, International Institute of Tropical Agriculture, Accra, Ghana. July 2011 version. pp 39-40.

Ananda, K.S. (2006) Cocoa Cultivation practices, Technical Bulletin, Central Plantation Crops Research Institute, Kasaragod, Kerala. pp 4-5.

4.4 CASHEW

Botanical Name	*Anacardium Occidentale*
Family	Anacardiaceae
Origin	South eastern, Brazil
Chromosome number	2n = 24, 30, 40,42
Commercial value	Kernels
Research Institute	National Research Centre for Cashew in Puttur, Karnataka

Introduction

Cashew has always been a neglected crop and in India, its large –scale cultivation is not possible due to enforced land ceiling laws of the state governments and till date it is not declared as a plantation crop. Although, it is an excellent crop for wasteland development and rocky areas in low elevation with minimum rainfall.

Cashew is a zero cholesterol nut and nutritionally it is at par with milk, eggs and meat. It contains 47% of fat, but 82% of this fat is unsaturated fatty acids. It reduces the cholesterol level. 21% protein and 22 % carbohydrate and a right combination of amino acids, minerals and vitamins make it nutritive cash crop. It does not lead to obesity and helps to control diabetes. It is good appetizer, an excellent nerve tonic, a steady stimulant and a body builder.

Distribution

India is the largest producer, consumer and exporter of cashew; mainly export to U.S.A., the Netherland, U.K. Germany, Japan, Australia and U.A.E. Major cashew growing countries are Brazil, Australia, Renin, Costa Rica, E1 Salvador, Guatemala, Honduras, Indonesia, Ivory Coast, Kenya, Madagascar, Malaysia, Mozarnbique, Nigeria, Panama, Philippines, Sri Lanka, Tanzania, Thailand, Togo, Venezuela, Vietnam, In India, Kerala, Maharashtra, Uttarakhand, Madhya Pradesh, Orissa, Karnataka, Tamil Nadu, Goa, and West Bengal are main producing state.

Description of plant

It is a low spreading, evergreen tree with a prominent tap root system, well developed and extensive network of lateral and sinker roots. Flowers either bisexual, or staminate or both. Perfect flowers are larger than staminate flowers. Under normal conditions 85% flowers are fertilized but only about 4-6% reaches at maturity. The fleshy peduncle the cashew apple is juicy and sweet when ripe. Rich source of vit. C and sugar. Cashew is a kidney shaped, drupaceous nut greenish grey. The weight of nut is 3-20g and 20-30% shelling.

Branching nature of Cashew:

i. Intensive

ii. Extensive.

Intensive shoots grow to a length of about 25-30cm and terminates in a panicle. Simultaneously, 3-8 laterals arise within 10-15 cm of apex giving bushy appearance. In extensive type of system- Shoots grow to a length of 20-30cm and rests. A bud sprouting 5-8 below the apex leads further growth. This continues for 2-3 years without any flowering. It produces a spreading tree. It is a highly cross pollinated.

Climate

It is a hardy tropical plant. It can grow up to 700m MSL. Rainfall of 50-250 cm with temperature tolerance of 25-50° C is suitable . It does not tolerate excessive shade and is suitable for places situated at 35° latitude on either side of the equator.

Soil

Cashew can be grown in almost all types of soils from sandy to laterite (upto an elevation of 600-700 m) including waste lands of low fertility. It grows and yields best in well drained red sandy loams and light coastal sands. Heavy clay soils, poor drainage conditions, very low temperature and frost are unsuitable. Sites prone to water logging and excessive alkalinity and salinity should be avoided.

Propagation

Seed propagation is seldom practiced now except to raise the rootstock material. Heavy seed nuts which sink in water are alone mixed with 2 parts of sand. They take normally 15-20 days for germination.

Vegetative Propagation Since cashew is a cross pollinated crop, vegetative propagation is recommended to produce planting materials that is true to type. One year old shoots as well as current season shoots are used for air layering and Soft wood grafting. Veneer grafting, Side grafting and patch budding are successful but nursery period extends up to 3-4 years.

Planting

Pit size of 45 × 45 × 45cm should be dug which is filled with Topsoil, Neem cake and FYM (10kg.). Planting should be done at a distance of 7m × 7m. Fortyfive days old seedlings should be transplanted in the prepared pit.

Intercrops

Tapioca, Groundnut and Pulses are used as suitable intercrop with cashewnut.

Training and Pruning

In older cashew plantations, removal of dried or dead wood, criss-cross branches, water shoots *etc.* should be removed at least once in 2-3 years. This allows proper growth of the canopy and receipt of adequate sunlight on all the branches. Pruning of cashew plants should be done during May / June.

Top working

Top working is a technique evolved to rejuvenate unproductive and senile cashew trees. Top working can successfully rejuvenate poor yielders in the age group of 5-20 years. The unproductive trees are to be beheaded at a height of 0.75-1m from ground level. The stem should be cut with a saw to avoid stump splitting. The best season for beheading trees is May-September. Soon after beheading, the stumps and cut portions should be given a swabbing with copper oxychloride. Sprouts emerge 30-45 days after beheading. Sprouting will be profuse in young trees. 20-25 days old shoots should be grafted with scions of high yielding varieties using softwood grafting technique. To ensure at least six or seven successful grafts, 10-15 grafting are to be done on the new shoots of every tree. The best season for grafting is July-November. Thinning of the extra shoots arising from the stumps should be done to obtain better growth of the grafts.

The top worked trees start yielding right from the second year after top working. Thus precocity can be considered as one of the best advantages of this technique. The major disadvantage associated with top working is the huge casualty of trees due to stem borer attack. Intensive care and management to ward off stem borer is essential. As such, adoption of top working on a larger scale would be difficult.

Manure and Fertilizers

Manures and fertilizers should be added as per schedule given below

Age of plant	FYM or compost (kg)	N(g)	P_2O_5(g)	K_2O(g)
1	10	50	25	25
2	20	100	50	50
3	20	150	75	75
4	30	150	75	75
5	50	500	125	125

Harvesting

February to May is the suitable time for harvesting.

Yield

Fruiting start from the 3rd year onwards and it continous upto 50 year. Yield 20 – 30 kg/plant can be obtained.

Plant Protection

Varieties

Varieties / hybrid / types	Mean yield (kg/tree/year)	Nature of bearing
Anakkayam-1 (BLA 139-1)	12.00	Early
Madakkathara-1 (BLA 39-4)	13.80	Early
Vridhachalam-3 (M 26/2)	11.68	Early
Kanaka (H-1598) (BLA 139-1 × H3-13)	12.80	Mid

Varieties / hybrid / types	Mean yield (kg/tree/year)	Nature of bearing
Dhana (H-1608) (ALGD 1-1 × K 30-1)	10.66	Mid
K-22-1	13.20	Mid
Dharasree (H-3-17) (T 30 × Brazil 18)	15.02	Mid
Priyanka (H-1591) (BLA-139-1 × K30-1)	16.90	Mid
Amrutha (H-1597) (BLA-139-1 × H3-13)	18.35	Mid
Anagha (H8-1) (T 20 × K30-1)	13.73	Mid
Akshaya (H7-6) (H4-7 × K30-1)	11.78	Mid
Madakkathara-2 (NDR 2-1)	17.00	Late
Sulabha (K10-2)	21.90	Late
Damodar (H1600) (BLA 139-1 × H3-113)	13.36	Mid
Raghav (H1610) (ALGD 1-1 × K 30-1)	14.65	Mid

4.5 COCONUT

Botanical Name	*Cocos nucifera* Linn
Family	Araceae.
Origin	South East Asia, most probably Malaysia or Indonesia
Chromosome number	2n=32
Commercial value	Nuts, Leaves
Research Institute	Central Plantation Crop Research Institute, Kerala

Introduction

Coconut is the only plantation crop, which allows maximum bio- diversity within its interspaces, maintaining ecological balance. The coconut palm is referred to as "*Kalpavriksha*" the tree of heaven, as each and every part of the palm is useful in one way or the other. It has varied uses as a source of food, drink, shelter and as raw material for various industrial uses. Immature nuts contain water (liquid endosperm), which has medicinal values of diabetic and heart patients. Its edible part is endosperm or meat, closely adhering to endocarp. The white meat or kernel is used in making chutney, curries, sauces, toffee, sweet etc. Dry copra is used for extraction of oil, it contains about 65-75% oil. Toddy, coconut cake, Jaggery *etc*. can also be obtained. Coir, which is used for making carpets and ropes are major products for export purposes. Leaf petioles and rachis or midribs are used for making brooms. Coconut shell is used as fuel. Coconut oil is the most important edible oil for domestic use. Coconut oil prevents skin diseases like eczema. It is world's most important oil crop.

Distribution

In India, Kerala has the longest history of coconut cultivation followed by Andaman and Nicobar, Lakshadweep and Goa. India has attained the top position in the production of coconut, overtaking Indonesia and Philippines.

Description of plants

Coconut palm is monocot. Palm has the unique feature of having only a single growing point on the whole tree. Once this growing point is damaged, the whole plant dies. The base of the trunk is thick, so called 'bole'. Its average life span is 70 years. Its phyllotaxy is alternate and two consecutive leaves are placed at an angular deflection of 137°. Due to the absence of outer cambium, they cannot increase in thickness. From the surface of main roots and rootlets, small pointed out growth develop, which function as the respiratory roots or pneumatophores. These roots do exchange of gases and make palm to stand water logging or marshy conditions. Inflorescence is monoecious; its form in the leaf axil and is enclosed in a double sheath or spathe, the whole structure being termed as 'spadix'. Inflorescence consists of main axis (peduncle with upto forty side branches (spikelets). These branches bear male flowers towards the apex and female flowers towards the base.

Coconut to highy cross-pollinated due to protendry. The trimerous syncarpous ovary usually produces one seed, other two ovules aborting, but the germ pore remaining at the micropylar end of the nut. Two pores are blind and the embryo lies below the third soft pore or eye. Polyembryony in coconut has been reported in the variety 'Arasikere Tall' in which a single nut gave rise three seedlings.

Cultivation

Climate

Coconut palm is heliophile (sun loving). A rainfall of about 2000 mm per year, well distributed throughout, is ideal for proper growth and maximum production. Mean annual temperature of 27°C with a diurnal variation of 5-7° C is ideal. Temperature <15°C results in abnormalities of the fruit. Coconut is predominantly grown under rainfed condition in Kerala and parts of coastal Karnataka and Tamil Nadu. In rest of the country, it is mainly grown under irrigated conditions.

Soil

Coconut is grown under different soil types such as loamy, laterite, coastal sandy, alluvial, clayey and reclaimed soils of the marshy low lands. The ideal soil conditions for better growth and performance of the palm are proper drainage, good water-holding capacity, presence of water table within 3m and absence of rock or any hard substratum within 2m of the surface.

Propagation

It can be propagated by seed and by tissue culture.

Dwarf types of coconut begin to form flowers 3-5 years after planting of seedlings, whereas tall types takes 6-8 years. Pollination is affected by bees. Mature fruit develop after 11-12 months of flowering. About 50-70% of the female flowers fall during button stage, due to failure of fertilization or other causes.

Selection of mother palm

a) The mother palms should be healthy, vigorous, regular bearers and should give annual yield of not less than 80 nuts.

b) Nuts should be medium in size, nearly round or spherical in shape having weight of more than 600 g and the copra content should not be less than 150g/nut under rainfed condition.

c) Selected palm should be 25-35 years of age if its parentage is unknown. Palms that have reached full bearing stage and giving regularly high yield for at least four consecutive years should be selected and very old palms above 60 years should be avoided.

d) Seed nuts can be collected from the newly established seed gardens irrespective of the age of the palms, since the performance of its parents are known and only high yielders are maintained in the seed gardens.

e) Palm should have atleast 30-40 fully opened leaves and 10-12 bunches. Petiole length and stalk of the bunches should be short and strong in nature. Long and thin petioles are not desirable because they are liable to be weak and may easily bend or break under pressure.

f) The trees with crown having leaf orientation to all directions (umbrella shape), spherical and semi-spherical crown is ideally suited as mother palms. Those trees having a drooping or upright position for the leaves are generally poor yielders and should be avoided.

g) Every leaf axil should have one inflorescence with a large number of spikes and one or two female flowers per spike. This ensures high nut-set and stable yield.

h) Mother palm should be free from insect, pest and diseases.

i) Selected palm should be growing very close to cattle houses.

j) Palms that produce barren nuts or those shedding large number of immature nuts and alternate/irregular bearers should be discarded.

Selection of seed nuts

a) Nuts should be selected from superior mother palms when attain full maturity (11-12 months). Always it is better to collect nuts that had undergone development during rainy season for seed nut purpose. Collect seed nuts from December to May.

b) Seed nuts should be round and oblong shaped without ridges.

c) Nuts should be larger in size. Discard nuts having irregular shape and size.

d) Should have greater copra content and give more percentage of germination. Avoid barren nuts.

e) Retain nut water for proper germination. Presence of water can be judged by hearing gurgling sound of water within the nut on shaking.

f) Fully matured nuts have dry husks with a distinct browning of the inner fibres, on tapping with finger ringing sound is produced. Immature nut produce dull sound.

g) Store seed nuts in shade for a minimum period of 60 days prior to sowing in nursery.

Nursery

Nursery sites should be well drained with light textured soil and with adequate but not too much shade. In open areas, provide shade during summer. Nursery can be raised either in the open with artificial shade or in gardens where the palms are tall and the ground is not completely shaded.

Prepare beds of 1.5 m width and of convenient length with 75 cm space between beds. In areas where drainage is poor, prepare raised beds. Sow the nuts in the nursery after commencement of southwest monsoon during May-June. Plant the seed nuts at a spacing of 30 cm (between rows) × 30 cm (between nuts) with four or five rows per bed.

Method of planting seed nuts

Plant the seed nuts in the beds in trenches 25-30 cm deep either vertically or horizontally and cover with soil so that top portion of husk alone is visible. The nuts are planted horizontally with the widest of the segments at the top or vertically with stalk-end up. Vertical planting is preferable on account of convenience in transporting and lesser risk of seedling injury. However, in delayed planting when the nut water goes down considerably it is good to go for horizontal sowing for better germination. Pre-treatment of seed-nuts at the time of planting is necessary. The nuts are examined individually and discard those nuts in which the nut water has dried up or kernel has been rotten in the course of storage. If the soil is dry and rains are delayed, soaking seed nuts in water prior to planting may be beneficial. Soak the nuts for about 48 hours.

Raising seedlings in polybags

In order to avoid the transplanting shock of the seedlings, it is possible to raise the seedlings in polybags. For this initially the nuts are sown vertically in the nursery closely and when they start germinating out in 8 to 10 weeks at every fifteen days the sprouted nuts are collected and placed in a black polybag of 500 guage measuring 60 cm x 45 cm, filled with potting mixture. The sprouts are thus collected for a period of 5 months from the date of sowing. These polybags are kept at a distance of 0.75 to 1.00 m from bag to bag.

Care and management of nursery

Provide protective fencing to the nursery if it is located in open area. If the soil is sandy, provide mulching immediately after the cessation of monsoon rain. Irrigate the nursery once in two days during summer months. Keep the nursery beds free of weeds by periodic weeding. Periodically spray the plants with 1 per cent bordeaux mixture or any other copper fungicide to prevent fungal infection.

Selection of seedlings

Remove seed nuts, which do not germinate within 6 months after sowing as well as those with dead sprouts. Generally 9 to 12 month old seedlings are used for planting. Select seedlings, which have 6-8 leaves and 10-12 cm collar girth when they are 9-12 month old. Early splitting of leaves (splitting should start atleast in the 8th leaf), early germination, rapid growth and seedling vigour are another criteria in the selection of coconut seedling.

Removal of seedlings from nursery

The seedlings should be removed from the nursery only when they are required for planting in the field. Remove seedlings from the nursery by lifting with spade and cutting the roots. Keep the seedlings in shade and do not expose to sun. Plant seedlings as early as possible after removal from nursery. Never allow lifting the seedlings from the soil by pulling the leaves or stem. The seedlings should be transplanted as early as possible though, under careful storage, they can be kept for nearly 4 weeks. The seedlings should be kept in shade and not exposed to the sun and also should be watered.

Land preparation and Planting

Shallow soils with underlying hard rock, low lying areas subject to water stagnation and clayey soils are to be avoided. Proper supply of moisture either through well distributed rainfall or through irrigation should be ensured before planting. The nature of preparation of land before planting depends upon topography of land, soil type and other environmental factors. The size of pits for planting would depend upon soil types and water table as given below:

Type of soil	Size of pit, m
Loamy soils with low water table	1×1×1
Laterite soils with underlying rock	1.2×1.2×1.2
Sandy soils	0.75×0.75×0.75

Spacing

It depends upon several factors *viz.*, soil, planting system etc. Coconut is an ideal crop for the introduction of multispecies cropping or mix cropping practices. A hectare of coconut plantation should not have more than 100-125 palms. In lateritic gravelly soils, under rainfed conditions, a closer spacing to accommodate 250 palms per ha is more economical.

Planting system	Spacing, m	Approximate number of plants per ha
Triangular	7.6 ×7.6	198
Square	7.6×7.6, 8×8, 9×9	170-220
Single hedge	6.5 in rows - 9 between rows	220
Double hedge	6.5 to 6.5 in rows - 9 between pairs of rows	280

Planting

Planting the seedlings during May, with the onset of pre-monsoon rains is ideal. Under assured irrigation, planting can be done during April also. In low-lying areas, plant the seedlings in September after the cessation of heavy rains. For the first two years from planting, irrigate @ 45 litres of water per seedling, once in 4 days, during dry summer months. Adequate shade should be provided to the transplanted seedlings.

Manures and fertilizers

To achieve higher productivity manures and fertilizers should be applied from first year of planting. For coconut 20 - 50kg organic manure should be applied per palm per year with the onset of south west monsoon, when soil moisture content is high. Different forms of organic manures like compost, farm yard manure, bone meal, fish meal, blood meal, neem cake, groundnut cake etc. could be made use for this purpose. Fertilizer schedule recommended is given in Table.

Under rainfed conditions

The dose recommended for the rainfed adult tall palm is 0.34:0.17:0.68kg NPK. For the hybrids and irrigated tall palm 0.5:0.34:1kg NPK recommended.

Under rainfed conditions, apply fertilizers in two split doses, 1/3rd at the time of early southwest monsoon showers in April-June and 2/3rd in September-October.

Under irrigated conditions, fertilizers should be applied i.e. in three or four equal doses in April-May, August-September, December and February-March. Apply lime or dolomite during April-May, magnesium sulphate during August-September and organic matter during June-July. For an adult palm 1.0 kg dolomite or 1.0 kg lime + 0.5 kg $MgSO_4$ is required per annum.

Apply fertilizers and manures in circular basins at a radius of 2.0 m from the base of the palm and 10 cm deep, opened after the onset of southwest monsoon. Split doses can be applied with irrigation water in summer months.

Irrigation

Coconut produces nuts round the year. Therefore, adequate supply of water is essential for its proper growth. Soil moisture very often limits coconut production in areas where long spell of dry weather prevail or where the rainfall is scanty and ill-distributed. So irrigate the palms during summer months in basins around the palm. The irrigation requirement varies according to the soil type and climatic condition. Generally, an adult palm requires 600 to 800 litres of water once in

Table: The fertilizer schedule recommended for the coconut palm at different stages is as follows:-

Age of Palm	Nutrient dosage	Quantity of fertilizer to be applied (gm)				
		Ammo. Sulphate	Urea	Single Super Phosphate	Rock Phosphate	Muriate of Potash
1. General recommendation:						
(a) Average Management:						
3 months	1/10 of full dose	165	75	95	60	115
1 year	1/3 of full dose	550	250	320	200	380
2 year	2/3 of full dose	1100	500	640	400	760
3 year onwards	full dose	1650	750	950	600	1140
(b) Good Management						
3 months	1/10 of full dose	250	110	180	115	200
1 year	1/3 of full dose	800	360	590	380	670
2 year	2/3 of full dose	1675	720	1180	760	1340
3 year onwards	full dose	2000	1080	1780	1140	2010
2. Hybrid and high yielding palms:						
a. Under Irrigated Condition:						
3 months	1/10 of full dose	490	220	280	180	335
1 year	1/3 of full dose	1625	720	930	600	1110
2 year	2/3 of full dose	3250	1450	1850	1200	2220
3 year onwards	full dose	4880	2170	2780	1800	3330

Source: http://www.srinidhifarms.com/tips_coconut

four to seven days. Irrigate in basins of 1.8m radius and 10-20 cm depth. In coastal sandy soils, sea water can be used for irrigating adult palms. Care should be taken that seedlings and very young palms upto 2 year should not be irrigated with sea water. In irrigated gardens interruption of irrigation would lead to serious set-back in yield and general condition of palms. Hence, when once started irrigation should be continued regularly and systematically. Drip irrigation is the best suited method of irrigation as it saves water, labour and energy.

Weeding and interculture

Keep the pits free of weeds by periodical weeding. Remove the soil covering the collar of seedlings. As the seedlings grow fill up the pits gradually by cutting the sides. Proper intercultivation provides control of weeds and creates soil mulch. Any tillage system (ploughing, digging, raking or forming mounds) that provides soil mulch and control weeds may be followed depending upon local conditions. For laterite, sandy and red sandy loam soils give two ploughings or diggings in May-June and September-October and one raking in January. In areas where surface run off is more, form mounds in September-October and level them in November-December.

Harvesting

Coconut ripen in 12-13 months from the opening of the inflorescence. When coconut water is required, the nuts of 5-6 months of age are harvested. When it has to be used for copra (culinary purpose) the nuts of 8-10 months are harvested. When dry copra is needed for coconut oil then fully ripe nuts of 10-12 months or so old are harvested. On an average, coconut yields 44 nuts/palm/year. However, under scientific cultivation West Coast Tall gives 80 nuts/palm/year in coastal Kerala and Karnataka. The hybrids yield 100-140 nuts/palm/year. Superior, golden-brown, quality fibre with elastic and good tungsten strength is obtained from 10-month-old nuts. The harvested nuts are stored in heaps under shade for a few days since the stored nuts are easy to husk. The moisture content of the meat decreases, whereas thickness of the meat layer increases. However, storage of harvested nuts is beneficial if fully matured nuts are harvested. Postharvest management of coconut involves its conversation into copra and coconut oil. Coconut husk is used to manufacture coir mat, cushion and other products.

Preservation of seed coconut

Preservation of seed coconut helps improper drying of the husk, which in turn helps to retain the nut water. When the nuts are not stored properly, the water inside may be lost and such dried nuts do not germinate. Sand storage of nuts practiced in most of the nurseries involves lot of labour and money. It has been proved that when the seed nuts are stored in cool dry places in thatched sheds or under the shade of trees practice of storing the nuts in the soil can be avoided. It may be necessary at times to sprinkle some water over the nuts so that the nut water inside does not dry up. It is preferable to store the nuts in separate lots based on their date of collection.

Plant protection

Insect- pests

1. Rhinoceros beetle (*Oryctes rhinoceros* L.)

It is popularly known as die back beetle and it is supposed to be the most serious in coconut growing regions. This pest affects seedlings, juveniles and adult bearing palms. Severe infestation of young plants leads to reduction in growth, sometimes even loss of such plants in the field. It can also induce considerable yield reduction by the damage inflicted in the unopened bunches in bearing palms.In few cases toppling of the spindle leaves may be noticed.

Control

i. Remove and burn all dead coconut trees in the garden (which are likely to serve as breeding ground) to maintain good sanitation.

ii. Collect and destroy the various bio-stages of the beetle from the manure pits (breeding ground of the pest) whenever manure is lifted from the pits.

iii. Soak castor cake at 1 kg in 5 litre of water in small mud pots and keep them in the coconut gardens to attract and kill the adults.

iv. Treat the longitudinally split tender coconut stem and green petiole of fronds with fresh toddy and keep them in the garden to attract and trap the beetles.

v. Examine the crowns of tree at every harvest and hook out and kill the adults.

vi. For seedlings, apply 3 Naphthalene balls/palm weighing 3.5 g each at the base of inter space in leaf sheath in the 3 inner most leaves of the crown once in 45 days.

vii. Set up light traps following the first rains in summer and monsoon period to attract and kill the adult beetles.

viii. Field release of Baculovirus inoculated adult rhinoceros beetle @ 15/ha reduces the leaf and crown damage caused by this beetle.

ix. Apply mixture of either neem seed powder + sand (1:2) @150 g per palm or neem seed kernel powder + sand (1:2) @150 g per palm in the base of the 3 inner most leaves in the crown.

x. Place phorate 10 G 5 g in perforated sachets in two inner most leaf axils for 2 times at 6 months intervals.

xi. Set up rhinolure pheromone trap @ 1/ 2 ha to trap and kill the beetles.

xii. Bio control methods: CPCRI, Regional Station, Kayamkulam has evolved effective bio control techniques for management of rhinoceros beetle, using green muscardine fungus (GMF), Metarhizium anisopliae. Rhinoceros beetle lays eggs in cow dung pits, vermicompost units, coirpith, degraded coconut logs etc. Treatment of such breeding sites with GMF effectively brings down the population of the pest. The fungus will not affect the earthworms in vermicomposts in any way. Different stages of grubs and adults will be infested by the fungus within a week of its application. Incorporation of *Clerodendron infortunatum,* a common weed seen in coconut gardens, in the breeding sites also inhibits the growth stages of the black beetle.

2. Black headed caterpillar (*Opesinea serinopa*)

It is commonly known as leaf-eating caterpillar and it is serious in coconut palm. The Symptoms of damage includes dried up patches on leaflets of the lower leaves. Galleries of silk and frass on under side of leaflets.

Control

i. The incidence of the pest is noticed from the month of November to May and from August to November after rainfall. The coconut trees of all ages are attacked.

ii. Release the larval (*Bethylid, Braconid* and *Ichneumonid*) and pupal (*Eulophid*) on (chalcid) parasitoids and predators periodically from January, to check the build up of the pest during summer.

iii. Biologically it is controlled by *Perisierolane phantidis*, *Microbracon brevicornis* and pupal parasite *Trichospilus pupivora.*

iv. Among the larval parasitoids, the bethylid *Goniozus nephantidis* is the most effective in controlling the pest. The optimum level of release is 1:8 of host-parasitoid ratio. The parasitoid should be released @3000/ha under the coconut trees when the pest is in the 2nd or 3rd instar larval stage.

Parasitoid release trap may be used to release the parasitoid at the site of feeding. Parasitoids should not be released in the crown region since they will be killed by predators like spiders and reduviid bugs.

v. Remove and burn all affected leaves/leaflets.

vi. Spray malathion 50 EC 0.05% (1ml/lit) to cover the undersurface of the leaves thoroughly in case of severe epidemic outbreak of the pest in young palms.

3. Red palm weevil (*Rhynchophorus ferrugineus* F.)

This pest is a serious as Rhinoceros beetle. Symptoms of damage are holes on trunk with brownish ooze. Yellowing of inner leaves. Gradual wilting of central shoot in the crown

Control

i. Remove and burn all wilting or damaged palms in coconut gardens to prevent further perpetuation of the pest.

ii. Avoid injuries on stems of palms as the wounds may serve as oviposition sites for the weevil. Fill all holes in the stem with cement.

iii. Avoid the cutting of green leaves. If needed, they should be cut about 120 cm away from the stem.

iv. Fill the crown and the axils of top most three leaves with a mixture of fine sand and neem seed powder or neem seed kernel powder (2:1) once in three months to prevent the attack of rhinoceros beetle damage in which the red palm weevil lays eggs.

v. Setting up of attractant traps (mud pots) containing sugarcane molasses 2½ kg or toddy 2½ litres + acetic acid 5 ml + yeast 5 g + longitudinally split tender coconut stem/logs of green petiole of leaves of 30 numbers in one acre to trap adult red palm weevils in large numbers.

Diseases

1. Leaf rot caused by fungus *Bipolaris haloides*. High humidity and low temperature conditions, which are common during rainy season have found conducive for rapid development of fungus.

2. Budrot caused by *Phytophthora palmivora* attack on growing point, which results in the death of palm. Disease is more prevalent in the gardens of low lying area.

3. Foot or trunk rot caused by fungus *Ganoderma lucidum,* it is soil and air borne disease. The roots and the base of the stem show discoloration and formation of bleeding patches at the base of trunks, dark brown liquid ooze out from the trunk.

   Drench Bordeaux mixture 2% or Bavistin 0.1% at 10-15 litre/palm at 30-40days interval in the month of June and continue upto September. Apply neem cake @10kg/palm/year.

4. Stem bleeding is another problem of coconut. Its cause is unknown, but it is thought to be associated *Thieoviopsis paradoxa*. It is a wood parasitic fungus. It invades through wounds, cuts, cracks into the palms. It can be controlled by pasting the cut portion with Bordeaux paste.

5. Root wilt and Eriophyid mite are menace recently causing damage to the coconut. It is a devastating disease prevalent in Kerala and it was observed in 1882.High water table and poor drainage conditions are supposed to be predisposing factors of the disease.

Physiological Disorder

1. Button shedding i.e. the immature nut fall is very common during the first two to four months after the opening of flowers. Maximum shedding is reported during 2nd and 3rd week after fertilization. This problem may be due to stigmatic non-receptivity, soil moisture stress, imbalanced availability of nutrients to the plant, infection of *Phytophthora maviora* fungus, and infestation of sucking insects and formation of abscission layer at the place of attachment to the stalk due to prevailing moisture stress conditions. This problem can be minimized by maintaining moisture. Apply adequate manure and fertilizers, spray any copper fungicides and growth regulators like 2, 4-D @ 60 ppm on female flowers at 7-10 days intervals just after completion of fertilization.

Varieties

Tall varieties

This is the ordinary or the common tall variety of palms most extensively grown on a plantation scale in all coconut tracts of the world. They are hardy, late bearing and lives upto an age of about 80 to 90 years. It has long and stout trunk with a swollen base called bole. This variety is characteristically tall, growing to a height of about 15 to 18 m. It begins to bear in about 8 to 10 years after planting. The palms of this variety are generally cross-pollinated in nature, although, in summer months there exist chances of self pollination due to the overlapping of spadices. After pollination, the nuts mature in a period of 12 months.

West Coast Tall (WCT)	This variety is known as the ordinary or common Tall variety. It is the one that is extensively cultivated in all the important coconut tracts of India and is of commercial importance. It is known as West Coast Tall since it's cultivation is mainly concentrated on the west coast of India. It is a long-lived, hardy, multipurpose palm, yielding nuts, copra, oil and fibre of good quality. The tree also yields, on tapping, good quantity and quality of coconut juice or toddy which can be fermented or converted into jaggery or sugar. Under favourable conditions the palms commence to yield in six to eight years after planting. The average yield under proper management and rainfed conditions is about 60 to 80 nuts per palm per year. The mean copra content is about 165-176g per nut and the oil content is 68-72 per cent by weight of copra.
East Coast Tall	It is a tall variety suitable for production of copra and tender nut. The palms are tall, compact and bear medium sized oval shaped green nuts. The average yield 14500 nuts/ha, copra yield 34 q/ha and oil 22 q/ha.
Kappadam	This type is popular in parts of South Malabar Region. It is essentially a tall palm with large sized nuts which are generally ellipsoid in shape. The yield is rather low being about 60 nuts per palm per year. Quantity of copra is high and copra is thick and hard. The out turn of copra per nut is about 285 grams. The quality of copra is good and is comparable to that of the west coast tall.
Andaman Giant	This is a type from the Andaman Islands. The palms are of tall stature, massive proportions and majestic appearance. The nuts are green and large, perhaps one of the largest on record but the yield is poor. Though the content is high, the copra is thin and of poor quality. It comes to bearing in about eight years after planting. The palms are resistant to diseases.

Dwarf varieties

The dwarf coconut is small in stature and commences bearing earlier than the tall variety. Dwarf coconut palms flower as early as the third year after planting and come to regular bearing in the ninth year, i.e. in about six years of first bearing. It has a thin trunk without a swollen base or 'bole' as compared to the tall type palms and its fully developed brand rarely exceeds 4 metre. Under normal conditions, the palm starts flowering in about three to four years when the trunk is just about the ground, and a fully grown tree rarely exceeds 5 m in height. It yields heavily, but has a tendency to irregular bearing. The Dwarf or short variety which has three distinct forms, namely, those producing green, orange and yellow nuts.

Chowghat Dwarf Green	It has dwarf-green coloured nuts and petioles, early flowering habit, i.e. three years after planting under ideal conditions, oblong nuts with tapering end - often they are edged with a beak at the end, the stem with a narrow base, and its girth is approximately 51 cm, compact crown with 20-26 short and delicate leaves; a fully grown leaf is 2.87 m long and 1.42 m wide with 0.91 m long petiole, overlapping male and female phases, large number of female flowers with spadix, and high annual yield per tree as compared to the ordinary tall. It maintains high purity, while in dwarf orange both self and cross pollination occur.
Chowghat Dwarf Orange	The stem with a small compact crown. Almost uniform size of the stem, close leaf scars on the stem, short stature, early flowering, i.e., three to four years after planting, short petiole, narrow leaf blades, closer arrangements of leaflets, overlapping male and female phases, retention of unfertilized female flowers, orange colour petiole, spathes and tender nuts, spherical orange colour mature nuts. Dwarf orange is slightly superior in bearing capacity to the Dwarf Green.
Andaman dwarf	This is a dwarf type from the Andaman Islands. It is a poor yielder. The copra content per nut is about 145 g and the oil content in the copra is about 63 per cent.

Gangabondam — This belongs to the 'medium dwarf' or 'semi-tall' type. It is found growing in some parts of Andhra Pradesh. The time of first bearing occurs in about four years after planting. The nut matures in eleven months. It is a prolific yielder. The nuts are elongated in shape, medium sized and have fairly thick kernel. They give good quality copra. The copra content per nut is about 230 g and oil percentage in the copra is about 72.

Hybrid palms

The manifestation of hybrid vigour in a perennial crop like coconut palm was first reported from India in 1932. The inter-varietal hybrids produced for commercial plantings are the Tall × Dwarf and Dwarf × Tall with different parental combinations.

Godavari Ganga — It is a hybrid between East Coastal Tall as female parent and Ganga Bondam as male parent. This variety was developed at Agricultural Research Station, Ambajipet. This variety is recommended for A.P and Tamil Nadu states. This variety starts bearing from 6th year and yields on an average of 140 fruits per tree. This hybrid contains copra of 15g and oil content of 68%.

Kalpa Sankara — This coconut hybrid (CGD x WCT) has tolerance to root (wilt) disease. The palms are semi tall in nature, precocious bearing and relatively higher yield in root (wilt) diseased tracts. The average yield 14700 nuts/ha, copra yield 25 q/ha and oil 17 q/ha.

Chandra Laksha — It is a hybrid (LCT x COD) comes to bearing in about 4-5 years after planting. The average yield 19075 nuts/ha, copra yield 37 q/ha and oil 26 q/ha.

Kera Sankara — The palm comes to bearing by the fourth year of planting. This is a T×D hybrid (WCT x COD). The average yield 18550 nuts/ha, copra yield 37 q/ha and oil 25 tons/ha.

VHC1 — This is a hybrid of East Coast Tall × Malayan Green Dwarf. The average yield 115 nuts/tree. The yield of copra 142 g/nut having 69% oil content.

Coconut Hybrid Mahuva	This hybrid was released from Junagadh Agricultural University, Junagadh, Gujarat. It is a hybrid of Tall X Dwarf. The average yield 43 nuts/tree.

4.6 ARECANUT

Botanical Name	*Areca catechu* L.
Family	Arecaceae
Origin	Malayan Archipelago, Philippines and other East Indian Islands
Chromosome number	2n=32 and is a secondary allotetraploid
Commercial value	Kernel
Chemical constituents	It has three alkaloids namely Arcoline, Arcolidine and Guracine.
Research Institute	Central Plantation Crops Research Institute (CPCRI), Kerala
Other institutes	CAMPCO (Central Arecanut Marketing and Processing Cooperative Ltd.), Mangalore.

Importance

The arecanut palm is the source of common chewing nut, popularly known as betel nut or Supari in India. It is an important cash crop obtained from an ordinary monocot plant. Arecanut forms an essential requisite for several religious and social ceremonies and its use dates back to Vedic period with high antiquities. Arecanut was even used in Ayurvedic and Ethane veterinary medicines. The practice of chewing the arecanut either alone or in combination with betel leaves of pan, lime, tobacco, camphor of spices, the combination has been in existence from time immemorial. Chewing is to increase the production of saliva and gastric juices and thus aid in the digestion. It is believed to strengthen the gums and the teeth along with cleansing and deodorizing the mouth. It is also an appetizer and a stimulant. The offering of betelnuts and flowers, placed on a few leaves of pan, in worship is a very common, traditional time honoured practice. It is used against leucoderma, leprosy, cough, fits, worms, anaemia and obesity, as a purgative. Tannins of Arecanut can be used in the leather industry or as an adhesive in the manufacture of ply board. The nut contains 8-12% fat that has characteristics comparable with hydrogenated coconut oil. It contains both saturated and unsaturated fatty acids. Arecanut fat can be extracted by using hexane as a solvent, and the fat can be made edible by refining it with an alkali. Simple blending of arecanut fat with butterfat followed by inter-esterification gives good products, acceptable in confectioneries. Fallen areca leaves and arecanut husk used for the cultivation of its economic produce is the fruit called 'betel nut' and is used mainly as masticator. Its organic waste could be successfully converted into fine granular vermicompost.

Distribution

Though it is native of Malayan Archipelago, Philippines and other East Indian Islands, Commercial cultivation is confined only in India, Bangladesh and Sri Lanka. Major states cultivating this crop in India are Karnataka (40%), Kerala (25%), Assam (20%), Tamil Nadu, Meghalaya and West Bengal. India also ranks first in the world in its area and production. India is also the largest consumer of Arecanut. High productivity of Arecanut is recorded in China followed by Malaysia. India has no competition and has good domestic market of arecanut. Its Value added products are exported to Gulf countries. It is exported to U.K., Arab countries and South East Asia.

Description of plant

The name *Areca* is derived from a Malayan word meaning "Clusters of nuts". It is an erect, unbranched monoecious palm reaching heights of 12-30 m, depending upon the environmental conditions. The stem, marked with scars of fallen leaves in a regular annulated form, becomes visible only when the palm is about 3 years old. Girth depends on genetic variation and soil conditions. The adult palm has 7-12 open leaves, each with a sheath, a rachis and leaflets. The leaf stalk extends as the midrib until the end of the leaf and ends as leaflets. Root system is adventitious. Inflorescence is a spadix and androgynous. Male flowers are many occupying the upper part of the spike while, female flowers are few at the base of spike. Fruit is monolocular, one-seeded berry, 3.8-5 cm long, smooth orange or scarlet when ripe, with a fibrous outer layer. It is a cross-pollinated crop. Male and female flowers are sessile.

Climate

The arecanut palm is capable of growing under a variety of climatic and soil conditions. The cultivation of arecanut is mostly confined to 28° north and south of the equator. It grows well within the temperature range of 14°C and 36°C and is adversely affected by temperatures below 10°C and above 40°C. Huge variations in temperature and diuarnal variations is not conducive for growth of palms. It grows well from almost sea level up to an altitude of 1000 metres in areas receiving abundant and well distributed rainfall or under irrigated conditions.

Soil

It can be grown successfully on laterite, red loam and alluvial soils. The soil should be deep (1.5m) and well drained. The palm does not withstand drought or water stagnation. The pH of the soil should be slightly acidic to neutral.

Propagation

It is done by seeds.

Being a perennial, seed-propagated crop, much care is to be bestowed on the production and selection of planting materials. The different stages in the production of quality planting materials include selection of mother palm, seed nuts and seedlings and adoption of proper nursery techniques.

Selection of mother palms: The mother palm selected should be based on the following criteria:

i. Earliness in bearing.

ii. High percentage of fruit set (more than 50%).

iii. Palms with shorter internodes, more number of leaves on the crown ie semi-tall to dwarf in stature.

iv. Producing at least 4 bunches in a year or high percentage of fruit set.

Selection of seed nuts: It depends upon:

i. Fully tree-ripen nuts from the middle portion of the middle bunch on the tree.

ii. Nuts of weight 35 g and above should be selected. Heavier nuts give higher germination percentage and produce vigorous seedlings. The nuts which float vertically with the calyx end pointing upwards when allowed to float on water are preferred.

iii. Discard nuts, which are undersized, malformed and low in weight.

Nursery techniques

Sow selected seed nuts soon after harvesting in sand beds prepared under shaded condition with the stalk end up and with a closer spacing of 5-6 cm. Cover the seed nuts with sand and irrigate daily. Germination starts in 40-45 days and continues up to 3 months. Transplant 90 days old sprout having 2-3 leaves into the secondary nursery.

Prepare secondary nursery beds with 150 cm width and of convenient length.

Apply Farmyard Manure (FYM) @ 5 tonnes/ha as basal dose. Transplant sprouts at a spacing of 30×30 cm at the onset on monsoon during May-June. Areca sprouts and seedlings are very delicate and do not withstand exposure to direct sunlight. Hence, provide partial shade by growing banana, ivy gourd or by raising artificial. Drainage should be good to avoid water stagnation. Provide irrigation during hot and dry months. Periodical weeding and mulching are necessary.

Seed nuts can also be sown in polythene bags (25×15 cm size ,150 gauge) after filling the bags with potting mixture containing 7 parts of loam or top soil, 3 parts of dried and powdered farm yard manure and 2 parts of sand.

Selection of seedlings: Twelve to eighteen-month-old seedlings are to be selected and transplanted in the main field. Seedlings with maximum number of leaves (five or above), minimum height and maximum girth are to be selected for planting. Plants with less than 20 cm girth one year after transplanting and less than 4 nodes 2 years after transplanting should be discarded. The selected seedlings should be removed with a ball of earth adhering to the roots for planting.

Field Planting: Select site with deep (not less than 2 m), well-drained soil, without high water table and provision for irrigation. Arecanut palm is very delicate and cannot withstand extremes of temperature and exposure to direct sun. Plant tall, quick-growing shaded trees on the southern and western sides of the plantation to provide protection from the sun –scorching. Soil depth and water table are important aspects to be considered while selecting site, since it determines the development of root system. In areas where South-West monsoon is severe, planting in the month of September-October is recommended. In other areas planting can also be done in the months of May-June. Spacing of 2.7×2.7m is followed.

The palms are highly susceptible for sun scorching. The seedlings should be given protection against the direct exposure to sun. This may be done either covering the plants with areca or coconut leaves or by raising crops like banana in between two rows of arecanut. Sun scorching is mostly seen during October – January. During this period even the stems of young palms have to be protected. For this a quick growing shade plant can be planted on Southern and Western sides of the garden.

Manures and Fertilizers

Adequate supplies of plant nutrients in the soil throughout the life of the crop are essential to get high yield. The recommended fertilizer dose is 100:40:140g NPK +20kg FYM. The fertilizers are to be applied in two split doses. One third of the fertilizer is applied in May - June and two third along with the organics during September-October. Fertilizers are applied in basins around the palm dug to a depth of 15-20 cm and 0.5-1.0 m radius leaving 20 cm from the base of the palm. After application, the soil is rolled up and covered with organic matter and soil.

Irrigation and Interculture

Irrigation should be done depending upon soil type and climatic conditions. It is done once in a week during November – December, in four days interval during March- April and daily during January – February. Adoption of drip irrigation mitigates the loss of water (upto 65%) through conveyance which occurs normally in flood irrigation. Drip irrigation also avoids excessive wetting of the root zone which is detrimental to the areca palm.

The initial period of five to six years is ideal for short duration crops while shade tolerant tree species can be grown as mixed crops later. Thus areca palm offers opportunities for temporal and special distributions of inter/companion crops eg Cocoa is grown as a mixed crop.

Keep the garden free of weeds and break up surface crust by light forking or digging after cessation of monsoon during October-November. In slopes, prevent soil erosion by terracing. Sow seeds of green manure-cum-cover crops such as *Mimosa invisa, Stylosanthes gracilis* and *Calapagonium muconoides* in April-May with the onset of pre-monsoon rains. Cut and apply them to the palms in September-October.

Plant Protection

Insect -pests

1. Mites

White mite *(Raoiella indica)* and Red mite (*Oligonychus indicus*)

They are sucking pest and suck sapfrom green portion of plant. This causes the leaves to turn yellow and bronzed appearance.

Control

Spray dimethoate or Dicofol at 0.05 per cent. Repeat spraying at 15-20 days interval if reoccurrence of pest occurs.

2. Spindle bud (*Carvalhoia arecae*)

They form colonies in the top most leaf axils and suck sap from tender spindle and leaves. The affected leaves show dry brown patches. Spindle fails to open. Severe infections leads to stunting of the palm.

Control: Placement of 2 g phorate 10G sachets on the top most two leaf axils prevents the pest attack during April.

3. Inflorescence caterpillar (*Tirathaba mundella*)

This is also a major pest. Caterpillar feed on the flowers and clamp the inflorescence into a wet mass of frass with silky threads.

Control: Force open the inflorescence out of the enclosing spathe and spray Malathion 50 EC (250 ml in 100 litres of water). Control slugs, which predispose inflorescence to the attack of caterpillar, by using bait of Metaldehyde.

Diseases

1. Koleroga (Mahali or fruit rot) (*Phytophthora arecae*)

It is the most severe disease of arecanut in high rainfall areas. The annual crop loss is about 75%. Water soaked lesions appearing on the nut surface near the perianth and spread over the other parts. Infected nuts shed without perianth.

Control

i. Spray Bordeaux mixture 1% as a prophylactic measure on all bunches three times in a year, one just before the onset of southwest monsoon and the rest at 40 days intervals. If monsoon season is prolonged give a third spray. Use rosin soda adhesive to ensure tenacity of the spray deposit on treated substrate.

ii. Remove and burn all fallen and infected nuts.

2. Basal stem rot (Anabe roga) (*Ganoderma lucidum*)

It occurs in ill drained and over crowded gardens. It is primarily soil borne disease but spread through air borne spores.

Control

i. Isolate affected palms by digging trenches 60 cm deep and 30 cm wide around, one metre away from the base and drench with Captan (0.3%), Calixin (0.1%) or Copper oxychloride (0.3%).

ii. Proper management of gardens is the best way of reducing the disease occurrence and better drainage management.

iii. Remove and destroy all severely affected palms and stumps of dead palms.

iv. Drench the soil with 1% Bordeaux mixture before planting healthy seedlings.

v. Discourage growing of collateral hosts of fungus such as *Delonix regia* and *Pongamia glabra* in the vicinity of gardens.

vi. Apply 2 kg neem cake per palm.

vii. Avoid flood irrigation and water flowing from infected palms to healthy palms.

3. Yellow leaf disease caused by mycoplasma like organism

It is a serious malady and upto 50% reduction in yield has been reported due to this disease. Leaves become yellow, smaller, stiff and pointed, crown gets reduced, and palm remains stunted with few or no nuts.

Control

i. Maintain the garden properly to keep affected palms in a healthy condition by adopting recommended manure application, cultural, plant protection and other management practices.

ii. Improve drainage conditions in the garden.

iii. Remove all the affected plants from the garden.

Varieties

Mangala	It is an introduction from China (VTL-3). It is high yielding semi tall palm with partially drooping crown, earliness in bearing, more number of female flowers/inflorescence, higher nut set, quicker stabilisation, round and medium sized yellow coloured nuts. The average yield 41 q/ha (dry kernel).
Sumangala	It is a selection from Indonesia (VTL-11). It is high yielding tall palm with partially drooping crown, oval to round shaped deep yellow coloured nuts. High recovery of chali (26.50%) from fresh fruits. The average yield 45 q/ha (dry kernel).

Sreemangala	It is a selection from Singapore (PTL-17). It is high yielding tall palm with sturdy stem, partially drooping crown, round and bold with deep yellow coloured nuts. The average yield 44 q/ha (dry kernel).
Swarnamangala (VTL-1m2)	Regular bearer, consistent yielder with homogenous population. It is high yielding tall palm with medium thick stem and comparatively shorter internodes, partially drooping crown with well placed bunches. Average no. of bunches/palm/year is 3-4. The colour of ripe nuts is Orange to deep yellow with Oblong to round shape and bold. It start bearing by 4th year. Potential yield (kg chali/palm/year) is 6.28. Average yield (kg chali/palm/year) - 3.88. High recovery with 26.52% of chali from fresh nut. Recommended regions/areas for cultivation- Irrigated areas of Karnataka and Kerala.
VTLAH-1	It is a hybrid variety (Hirehalli dwarf x Sumangala). It is dwarf in nature. Sturdy stem with super imposed nodes, reduced canopy size, well spread leaves, medium sized oval to round shaped nuts and early stabilization and medium yielder. The main advantages of the hybrid variety is reduced cost of cultivation in terms of harvesting and spraying. The average yield 35 q/ha (dry kernel).
VTLAH-2	It is a hybrid variety (Hirehalli dwarf x Mohitnagar). It is dwarf in nature. Sturdy stem with super imposed nodes, reduced canopy size, well spread leaves, medium sized oval to round shaped nuts and early stabilization are the salient features of this hybrid variety. It is a medium yielder. The main advantages of this hybrid variety is reduced cost of cultivation in terms of harvesting and spraying. The average yield 36 q/ha (dry kernel).
Hirachelli dwarf	It is a natural dwarf mutant of arecanut.

Reference

Gurung, T, (Ed.). (2007). HC21: Arecanut Handouts. Lobesa: College of Natural Resources.

4.7: BETELVINE

Botanical Name	*Piper betle* L.
Family	Piperaceae
Origin	Central and Eastern Malaysia
Chromosome number	x=13 (Basic)2n=26,52 (Samuel and Bavappa, 1981)2n=32,64 (Dasgupta and Datta,1976) 2n=78 (Jose and Sharma, 1988)
Commercial value	Leaves

Importance

Betel vine commonly known as "*pan*" is a perennial creeper and cultivated in India for its leaves since time immemorial. Its deep heart shaped leaves are used for chewing and other parts have medicinal uses. It occupies an important place in everyday life of Indians. It is considered auspicious to make offerings with the betel leaf and arecanut on occasion such as wedding and worships. Betel leaf is known traditionally to be used for the treatment of various diseases. The essential oils present in the leaves are known to possess an anti bacterial, anti protozoan and anti fungal properties. Therefore, it helps in control of bad breath, boils, conjunctivitis, constipation, headache, hysteria, itches, ringworm, swelling of gum, rheumatism, typhoid, cholera and tuberculosis etc. Its roots are known to have female contraceptive effects. Pan is also believed to provide strength to heart and regulate blood. In ayurveda it acts as *vata* and *kapha* suppressant. It also helps in expelling out the mucus from the respiratory tract because of its hot potency (Balkrishna, 2008). According to Yunani system the leaf has a sharp taste and good smell improves taste and appetite, tonic to brain, heart and liver, lessens the thirst, clear the throat and purify the blood. The harmful effects of pan as described in the *Ayurvedic* texts are that it weakens teeth, impairs health and deadens the taste buds of the tongue. In the Indian subcontinent, where chewing tobacco with pan is a common habit, cancer of the mouth is very common.

Phyto-constituents: The leaf contains Water (85-90%), Proteins (3-3.5%), Carbohydrates (0.5-6.1%), Minerals (2.3-3.3%), Fat (0.4-1%), Fibre (2.3%), Essential oil (0.08-0.2%), Tannin (0.1-1.3%), Alkaloid (arakene). It also contains different vitamins like Vitamin-C (0.005-0.01%), Nicotinic acid (0.63-0.89mg/100g), Vitamin-A (1.9-2.9mg/100g), Thiamine (10-70μg/100g), Riboflavin (1.9-30μg/100g) beside this it contains minerals such as Calcium (0.2-0.5%), Iron (0.005-0.007), Iodine (3.4μg/100g), Phosphorus (0.05-0.6%), Potassium (1.1-4.6%) [Guha, 2006]. Leaves contain bitter compounds that are about

(0.7-2.6%). The specific strong pungent aromatic flavour in leaves is due to phenol and terpene like bodies (Bajpai *et al.,* 2007). The total phenol content vary according to gender. The male plant contains three fold higher total phenols content and two fold higher thiocyanate content as compare to female plant. The quality of the leaf depends upon the phenolic content, i.e., more the phenolic content better the leaf quality (Bissa *et al.,* 2007).

A betel leaf garden once established is like a perennial fountain to the farmers, providing cash for meeting their everyday requirement. Betelvine cultivation is highly intensive and it especially suits small holdings. The leaves are also in great demand in several other countries of the world where it is either not grown at all or the demand exceeds the local supply. They are exported to countries like Bahrain, Canada, Great Britain, Hong Kong, Kuwait, Nepal, Pakistan, Saudi Arabia and many other European countries.

Distribution

Betel vine is a native of Central and Eastern Malaysia from where it has spread throughout Tropical Asia and Malaysian Madagaskar and East Africa at the later date. Major betel growing countries are Sri Lanka, India, Thailand and Bangladesh. Betel vine is widely distributed in all over the state of India except northern regions (Jammu and Kashmir, Haryana, Punjab, Himachal Pradesh), due to sever winter and northern west (Rajasthan, Gujarat) due to hot dry summer. In India it is grown as a cash crop in the southern parts, mainly in Andhra Pradesh, Karnataka, Kerala and Tamil Nadu. It is also cultivated in Bihar, Madhya Pradesh, Maharashtra, Orissa, Uttar Pradesh and West Bengal.

Description of plant

It is a perennial, dioecious, ever green climber. Woody climber with adventitious roots at swollen nodes. Leaf simple, alternate, cordate, 8-12 cm wide, 12-16 cm long, with description odour and spicy taste. Inflorescence in axillary spike; flowers unisexual, white. Fruit globose berry.

Climate

Usually the male plant are cultivated throughout India except the dry Northeastern parts. In Indian condition the female hardly produce flowers or fruits. In spite of that vines are cultivated for harvesting the heart shaped green leaves. Tropical, highly humid with 40-80% relative humadity, high rainfall (2250-4750 mm) and a shady places are suitable for vigorous growth. Hot dry winds are harmful and retard the growth of the vine.

Soil

It can be grown under heavy clayey loam, laterite and sandy loam soil. Soil with good organic matter and drainage system is best suited for betel vine growth. However, it can be grown on different types of soils such as heavy clayey loam, laterite and sandy loam soils. It does not tolerate saline and alkaline conditions.

Propagation

Betel is usually propagated by using stem cuttings. Cuttings should be healthy and taken from mother vines with high yielding and leaves are comparatively bigger, dark green in colour with 3-5 nodes. Cuttings should be taken from the apical and middle portion of the vines. Cuttings can directly be field planted or can be planted as rooted cuttings established in poly bags filled with a mixture of equal parts of top soil, cow dung, coir dust and sand by burying 2-3 nodes in the soil.

Cultivation system

There are two system followed in the *pan* cultivation

i. Open system- In this system cultivation using support plants. Support plants are raised for natural support and shade. These plants are sown in 45-60cm rows at least 45 days before planting the cutting of the leaves. The plants used are *Sesbania grandiflora, S. sesbane, Erythrina variegate, Moringa oleifera.*

ii. Closed system- Using artificial rectangular structures called *bareja* a rectangular structures for artificial support and shade. This practice of betelvine cultivation is older than 600-400 B.C. *Barejas* are made up of locally available materials such as wooden poles, bamboos stalk and a variety of grasses as thatching material. This small hut like structure are rectangular in shape and of approximately 2-2.5m in height and 0.02 ha in area. The normal workable size is 50×30m. It is made on slightly sloppy land, with the ensured irrigation source. It is compulsory to keep the east and west side shorter than north and south sides. This shape helps in humidity and temperature control. Westerly winds are not only dry but also high or low temperature depending on the season. In order to minimize the wind impact, thatching on the west side is thickest so that the wind velocity entering the *bareja* is reduced. Thatching on the north side of the *bareja* is thicker than on the eastern side. The side facing south is kept thinnest, possibly to allow some light from that direction. Site having proper drainage in all directions.

Soil Preparation

Soil should be prepared well by 4-5 ploughing and land should be raised by 5-10cm from the adjacent areas, provide gradient on both sides for quick drainage. Afterwards, field beds are prepared of 15cm high and 30cm broad. Before planting the cuttings, soil should be sterilized.

Soil Sterilization

For new plantation- application of Carbofuran @ 1.5kg/ha or neem cake (0.5t/ha)+ Carbofuran @ 0.75kg/ha. Carbofuran should not be recommended in established gardens at any stage.

Required cuttings: 40,000–75,000 per hectare for open system

1,00,000–1,20,000 for closed (*Bareja*) system

Planting time

For Betelvine cultivation

June-July under closed system while October month is best time under open system. Planting seasons and spacing varies from state to state which are as follows:

Table: Planting Time and Spacing in Different States

State	Planting season	Spacing
West Bengal	June- July and September- October	R-R=50-70cm,P-P=10-20cm
Uttar Pradesh	October- November	R-R=100cm,P-P=10-15cm
Maharashtra	July- August and October- November	R-R=80cm,P-P=20cm
Bihar	June-July, September and May- June	R-R=80-100cm,P-P=10-20cm
Madhya Pradesh	January-March	R-R=50-60cm,P-P=15cm

Training and Pruning

One month after planting, young shoots start appearing and these young shoots are trained along the support and tied with them using banana fiber or jute fiber once in every 15 to 20 days.

Lowering of vine

Betel vine has two very short periods of active growth, from February to mid-April and from July to mid-November. But the growth is greatly reduced due to extreme temperature between summer and winter. To cope up with the extreme summer heat the vines are lowered to the height of 1 m or less (above ground) by tilting at an angle of 40-45°. This reduces desiccation of leaves due to hot air.

Manures and Fertilizer

Oil seed cakes like castor cake, linseed cake, sesamum cake or neem cake are applied as manure @15q/ha. Nitrogen @ 200kg/ha/year as farmyard manure or oil cakes should be applied. A dose of 100kg each of P_2O_5 and K_2O/ha/year is also recommended. The fertilizers should be applied in 4-6 split doses at 2-3 months intervals.

Irrigation

It requires high soil moisture; provide frequently light irrigation. Proper drainage is essential during rainy season.

Harvesting

In about 3-6 months time, vines grow to a height 150-180 cm leaves are harvested. Harvesting starts from the lower portion of the stem. Leaves are removed along with the petiole with the right thumb. Once harvesting is commenced, it is continued almost every day or week. The interval of harvesting varies from 15 days to about a month till the next lowering of vines. After each harvest, manuring has to be done. Different states having different time of harvesting which are as follows:

Time of harvesting in different states

Name of state	Time of harvesting
Uttar Pradesh, Madhya Pradesh and Bihar	March-April
Andhra Pradesh	May- June
Tamil Nadu	January- February or April- May

Yield

About 75 to 100 lakh leaves/ha/year can be obtained.

Plant protection

Insect-Pest

I. **Scale insects**

Control

1. Select scale-free seed vines.
2. Spray Chlorpyriphos 20 EC 2 ml/lit when one or two scales are noticed on the basal portion of the stem/leaves.
3. Direct the spray solution to the basal portion of the vines. Spray NSKE 5 % or Malathion 50 EC 1 ml/lit.

II. **Mites (Sevvattai):** Mites can be controlled by spraying Wettable sulphur 50 WP @ 1 g/lit

III. **Sooty mould (Aphids):** To control aphids spray Chlorpyriphos at 2 ml/lit on Agathi leaves. Clip off excess Agathi leaves.

IV. **Mealy bugs:** Mealy bugs can be controlled by spraying Chlorpyriphos 20 EC at 2 ml/lit or Dimethoate 30 EC 2ml/lit.

Diseases

I. Phytophthora Wilt

Integrated method is applied for the management of *Phytophthora* wilt.

Control

1. Select well matured (more than 1 year old) seed vines free from pest and diseases.
2. Soak the seed vines for about 30 minutes in Streptocyline 500 ppm or Bordeaux mixture 0.5 %.
3. Apply 150 kg N/ha/year through Neem cake (75 kg N) and Urea (75 kg N) and 100 kg P_2O_5 through Super phosphate and 30 kg Muriate of potash in 3 split doses first at 15 days after lifting the vines and second and third dose at 40 - 45 days intervals. Apply on beds, shade dried neem leaf or *Calotropis* leaves at 2 t/ha and cover it with mud (2 t in 2 split doses).

II. Bacterial leaf spot, blight and bacterial stem rot

Control

1. Spray Streptocyclin @ 400 ppm + Bordeaux mixture @ 0.25% at the time of first disease symptoms appear.
2. Continue spraying at 20 days intervals. Always spray the chemical after plucking the leaves.

Varieties

Based on shape, size, brittleness, and taste of leaf blade, betelvine is classified into pungent and non-pungent varieties.

S. No	State	Popular Varieties
1.	Andhra Pradesh	Karapaku, Chennor, Tellaku, Bangla and Kalli Patti.
2.	Assam	Assam Patti, Awani *pan*, Bangla and Khasi *Pan*.
3.	Bihar	Desi *Pan*, Calcutta, Paton, Maghai and Bangla.
4.	Karnataka	Kariyale, Mysoreale and Ambadiale.
5.	Kerala	Nadan, Kalkodi and Puthukodi.
6.	Madhya Pradesh	Desi Bangla Calcutta and Deswari.
7.	Maharashtra	Kallipatti, Kapoori and Bangla (Ramtek).
8.	Orissa	Godi Bangla, Nova Cuttak, Sanchi and Birkoli.
9.	Tamil Nadu	Pachai Kodi, Thulasi, Venmani, Arikodi, Kalkodi, Karilanchi, Karpuram, Chelanthikarpuram, Koottakkodinandan, Perumkodi, Amaravila and Pramuttan, Kallarkodi, Revesi, Karpuri, SGM 1, Vellaikodi, Pachaikodi, Sirugamani 1, Anthiyur kodi, Kanyur kodi.
10.	Uttar Pradesh	Dealwari, Kapoori, Maghai and Bangla.
11.	West Bengal	Bangla, Sanchi, Mitha, Kali Bangla and Simurali Bangla.

References

Bajpai V., Sharma D., Kumar B., Madhusudanan K.P. (2010). "Profiling of Piper betle Linn. Cultivars by Direct Analysis in Real Time Mass Spectrometric Technique", *Biomedical Chromatography*, 24(12), pp. 1283-1286.

Balkrishna A. Secrets of Indian Herbs for Good Health. Divya Prakashan, Uttarakhanda, (2008). pp. 32.

Bissa S., Bohra A., Songara D. (2007). "Traditions in Oral Hygiene: Chewing of Betel (*Piper betle* L.) Leaves", Scientific Correspondence, 92(1): 24-28.

Guha P. (2006). "Betel Leaf: The Neglected Green Gold of India", *J. Hum. Ecol.*, 19(2): 87-93.

http://openmed.nic.in/2055/01/Siddha_herbs. pdf. 21 November, 2012.

Jose J and Sharma AK (1988) Karyomorphological Studies on Piper betel L. *Acta Botanica Indica* 16:186–194.

Dasgupta A.and Datta PC (1976). Cytotaxonomy of Piperacee.*Cytologia* 41:697–704

Samuel MRA and Bavappa KVA (1981) Chromosome Number Ingenus Piper. *Curr Sci* 50:197–198.

5

Production Technology of Medicinal Crops

5.1 AJWAIN/ BISHOP'S WEED

Botanical Name	*Trachyspermum ammi* (L.) Sprague
Family	Apiaceae (Umbelliferae)
Origin	The Middle East, possibly in Egypt
Commercial value	Fruit, oil from seeds
Chemical constituents	Thymol (from seed oil)

Importance

Ajwain is an important seed spice and is also useful for its medicinal values. It is a traditional potential herb for curing many ailments in human and animals. It is usually used as a digestive aid, relieves abdominal discomfort due to indigestion and antiseptic. The seeds of ajwain is bitter, pungent and contain 2-4.4% brown coloured oil known as ajwain oil. The main component of this oil is Thymol, which is used as gastrointestinal ailment, lack of appetite and bronchial problems. Thymol is also used in toothpaste and perfumery (Joshi, 2000). Seeds are also used for curing abdominal pains, abdominal tumors and piles (Krishnamoorty and Madalgeri, 1999). In southern parts of India dry ajwain seeds are powdered and soaked in milk, which is then filtered and fed to babies as it relieves colic in babies. It can be used as digestive mixture in large animals.

Distribution

It was originated in the Eastern Mediterranean region, probably in Egypt, and came to India with the Greeks, who were called Yavanas by South Indian. It is cultivated in Iraq, Iran, Afghanistan, Pakistan and India. In India it is cultivated in Rajasthan, Gujarat, Madhya Pradesh, Uttar Pradesh, Maharashtra, Bihar and West Bengal where Rajasthan produces about 90% of India's total production.

Botany

Ajwain is a cross pollinated crop. Flowers are self fertile, but cross pollination occurs through insects. Plant parts usually consumed are herb,volatile oil, and seeds.

Description of plant

It is a profusely branched, erect, glabrous or minutely pubescent annual herb, 60-90 cm tall. The stems of which are straite, the leaves are distant, 2-3-pinnately divided with linear segments. The flowers occur in terminal or seemingly-lateral pedunculate, inflorescence is compound umbels with 16 umbellets each containing upto 16 flowers. Flowers are actinomorphic, white in colour and small. They are male and bisexual with 5 corolla, bilobed petals, 5 stamens alternating with petals, inferior ovary and knob like stigma. The fruits are aromatic ovoid, muricate, cremocarps, greyish brown with persistant stylopodium. The fruit consist of two pericarps (which are the components of the fruit), greyish brown ovoid, compressed about 2 mm long and 1.7 mm wide, 5 ridges and 6 vittae in each pericarp, usually separated by 5 primary ridges and is 1-seeded.

Soil

It is widely grown in arid and semi arid regions where soil contain high levels of salt. The crop can be grown on wide variety of soil from heavy clay to elite loams. It is very sensitive to water logging and need good drainage. It can tolerant drought.

Climate and soil

This crop grows well in eastern India. It can be grown under wide variety of soil from heavy clay to elite loams.The crop is sensitive to waterlogging hence drainage is required. It can tolerate drought.

Propagation

It is propagated through seeds.

Sowing time

The best time for sowing of this crop is early August.

Seed rate

1.5 kg of seeds per hectare is required for sowing.The seeds are sown directly in the field. The seeds take 8 to 10 days to geminate. Thinning should be carried out when the plant gets a height of 20 to 25 cm.

Manuring and Fertilization

About 25 t/ha farmyard manure should be applied at the time of land preparation. Phosphate @ 30kg and Potash @ 30kg per hectare is incorporated into the soil at the time of sowing. Urea is applied in three equal split doses. First at the time of sowing, then subsequently two doses being at an interval of 30 days.

Harvesting

The plant requires about 120 to 140 days for maturity. When seeds turn full brownish the whole plant can be uprooted. The whole plants are then tied in small bundles and tacked for drying and then threshed.

Yield

It varies from varieties, soil and climatic conditions. An average of 184 umbel/ plant, 16-17 seeds/umbellates, and 17-18 umbellates/umbel is obtained. Seed yield is about 11-12q/ha with 2.01g test weight.

Plant Protection

Sucking pests including aphids and thrips were observed in moderate to severe form. Wilt, blight infests the crop and yield is affected.

Varieties - AA1, AA2, AA93, GA1, Lam sel 1

Sl. No.	Particulars	Remarks
1.	Name of the variety	• **Ajmer Ajwain 1**
•	Name of the Institute/s (Collaboration/projects)	National Research Centre on Seed Spices, Tabiji, Ajmer (Raj.).
•	IP Status	Ready for commercialization
•	Characteristics and special features	• It is suitable for cultivation both under irrigated and rainfed conditions. • The average height of plant is 112 cm. • It bears average number of 219 umbels / plant. • It is a late variety and takes about 165 days to maturity. • Seed yield 11-14 quintals /ha. • The seeds yield essential oil content of 3.4%.
•	Suitable growing region and season	• It is suitable for cultivation both under irrigated and rainfed conditions. During rabi as well as kharif season.
2.	Name of the variety	• **Ajmer Ajwain 2** (Ajwain variety for rainfed cultivation)
•	Name of the Institute/s (Collaboration/projects)	National Research Centre on Seed Spices, Tabiji, Ajmer (Raj.).
•	IP Status	Ready for commercialization
•	Characteristics and special features	• It is suitable for cultivation under rainfed conditions. • The average height of plant is 80 cm. • It bears average number of 185 umbels / plant. • It matures in about 147 days. • Seed yield 10-12 quintals /ha. • The seeds yield essential oil content of 3.0%.
•	Suitable growing region and season	It is suitable for cultivation both under irrigated and rainfed conditions. During rabi as well as kharif season

References

Joshi, SG 2000. Medicinal Plants. 1st Ed. Delhi (India) Oxford and IBH Publisher.

Krishnamoorthy, V. and Madalageri,M.B. 1999. Bishop weed *Trachyspermum ammi* (L.) an essential crop for north Karnataka. *Journal of Medicinal and Aromatic Plant Sciences* 21 (4):996-998.

5.2 DILL

Botanical Name	*Anethum graveolens*
Family	Asteraceae
Origin	Mediterranean region (Egypt)
Chromosome number	-
Commercial value	Dill seed (the fruit of the plant) Dill weed (the top eight inches of the leaves)
Chemical constituents	Limonene, Carvone, and Phellandrene
Research Institute	National Research Centre on Seed Spices, Ajmer

Importance

The plant yields two different herbs: dill seed (the fruit of the plant) and dill weed (the top eight inches of the leaves). Its leaves are used fresh or dried as an herb in dips, soups, salads, and other dishes. The seeds are used as a spice for pickling and for adding flavour to stews and roasts. The pale, yellow oil of dill is mainly composed of limonene, carvone, and phellandrene. Carvone is responsible for the strong and sometimes acrid flavour of dill and it is also a known antiflatulent. Limonene and phellandrene can cause skin problems such as irritation or sensitivity to the sun. Handling dill plants may cause contact dermatitis or make your skin more sensitive to sunlight. Both reactions are most likely to occur among food workers who handle dill. This is due to Limonene and phellandrene found in oil of dill and many other essential oils are photosensitizers.

Dill is frequently used in Ayurvedic and Unani medicines for indigestion, fevers, ulcers, uterine pains and kidney and eye problems. It is carminative, stimulant, stomachic, alterative, nervine, promotes milk flow, antispasmodic, emmenagogue, diuretic, galactagogue. It is constituted of essential oils include carvone, anethofuran, alpha-phellandrene, limonene, phellandrine, coumarins, glycoside in seeds. The leaves also contains vitamin C and minerals. The herb helps relax the smooth muscles of the digestive tract. Historically, injured knights were said to have placed burned dill seeds on their open wounds to speed healing.

It is believed to have its beginnings in the Mediterranean region. The plant has a long and ancient history in many countries as a culinary and medicinal herb. The earliest known record of dill as a medicinal herb was found in Egypt 5,000 years ago, when the plant was referred to as a "soothing medicine." Dill seeds are often called "meeting house seeds" because they were chewed during long church services to keep members awake or kids quiet. The seeds were also chewed in order to freshen the breath and quiet noisy stomachs.

Distribution

In India it is grown in the states of Rajasthan, Gujarat, Jammu & Kashmir, Uttar Pradesh, Odisha, Madhya Pradesh and Punjab.

Botany

Anethum sowa, Indian dill, resembles the European plant, but the seeds are longer and narrower, the ridges are paler, and they taste slightly different.

Description of Plant

An erect aromatic hardy annual (sometimes also grown as biennial) of the parsley, cilantro, fennel, family. Depending upon the variety, the plant can grow from 30cm (for dwarf varieties) up to 120 cm in ideal conditions. It produces a single tap root that can reach deeply into the ground in ideal conditions. The branching stems have white to off-white, vertical striations that run down their length, and they are devoid of hair. The leaves are thin, wispy and fern like. Unlike fennel, which resembles dill in appearance, the stems of dill are hollow. They also end with a broad, expansive inflorescence that is topped with yellow, scented, hermaphroditic flowers. Dill's inflorescences are arranged in umbels, in which the flower stalks develop from a common point. Flowers are yellow in colour and develop into dry fruits called schizocarps that are split into two parts, each one holding seeds. The seeds are oval and progress from a bright green colour when they first develop to dark brown later in the season.

Soil

Dill can grow fairly well in poor soil conditions. But it grows well in a light to medium- textured sandy or loamy soil that is kept moist and has adequate drainage. The plant also responds well to many types of organic soils. The recommended pH range of the soil is 5.0 to 8.2, with an average of 6.5. The soil temperature should remain at about 20-21°C.

Climate

Dill is considered as a cool weather crop and it grows well in temperatures ranging from 6-26° C. It should be planted in an area that receives 6 to 8 hours of direct sunlight every day. High winds can cause great damage to the crop because the hollow stems break and bend easily. Staking may help minimize the damage. It grows best in full sun. Hail, low moisture and hot weather have a detrimental effect on this herb.

Cultivation

Sowing time

Dill is primarily a summer season crop in temperate regions. It requires a cool and dry weather. In Northern plains it is sown during *Rabi* season (October-November).

Propagation

It is propagated through seeds.

Seed rate and Sowing

Seed rate of dill varies according to the growing condition. About 3 kh/ha is required for irrigated crop whereas 5 kg/ha for crop grown in rainfed condition.

Seeds are sown 1.9 to 2.5 cm deep, allowing 30 to 38 cm between plants and 45 cm between rows. Stagger plantings by 2 to 3 weeks for a continuous harvest. Seeds usually germinate in 21-25 days depending on the soil temperature. It needs light to germinate.

Manuring and Fertilization

The recommended dose of fertilizers for dill includes 90 kg nitrogen and 50 kg P_2O_5 per ha which may be supplied through urea and DAP, respectively. Full quantity of phosphorus and half of the nitrogen was applied as basal dose at the time of sowing and the remaining nitrogen was applied 30 days after sowing.

Intercultural operations

Dill is a hardy plant and can tolerate temperatures down to -3.8°C (25°F) once established. Dill plants require little water, usually irrigating once or twice a week is sufficient and one light application of fertilizer should be enough to sustain the plant for the entire growing season. If the plant is not sufficiently sheltered from wind, the plants will benefit from staking to prevent the hollow stems from snapping. If continuous supply of dill leaves are required then bolting should be avoided. This could be done by keeping the tops trimmed at regular intervals.

Harvesting

Dill is ready to harvest approximately 90 days after planting. Dill weed is best harvested before the plant is fully mature and before the flower buds have opened. Fresh cut dill can be stored in the refrigerator safely for two to three days. The stems can be placed in a cup of water to help keep the leaves fresh. Seedheads should be harvested 2-3 weeks after bloom before the seeds begin to change colour. The seedheads can be hung up to finish drying.

The optimum time for harvesting dill is early morning and evening. The higher moisture content of the plants when harvested at this time results in better flavour and the possibility of seed shattering is minimized.

Collection of Seeds

Dill seed can be easily collected by hand using the following method: First, place a brown paper bag over the seed heads and tie the opening closed. Cut the stem off at the base of the plant. Next, hang the stems upside down in a warm, well ventilated area to dry. Take the stems down after about two weeks and crush the dried seed heads in your hands over a container to separate the seeds from the seed head. An additional method involves laying the freshly harvested seed heads on a cookie sheet and then placing them in the freezer. Remove the frozen seed heads after a few days and rub the seed heads between your hands over a piece of paper to harvest.

Yield

10-15q/ha seed yield under irrigated condition and 6-7q/ha for rainfed condition. Essential oil 0.9-1.5% from herbs and 2.5-4% from seeds are obtained.

Plant protection

Insect-pests

1. Root knot nematode *Meloidogyne* spp

Due to nematodes Galls are developed on roots which can be up to 3.3 cm in diameter but are usually smaller. This leads to reduction in plant vigour and yellowing of plants. Galls can appear as quickly as a month prior to planting;

Nematodes prefer sandy soils and damage in areas of field or garden with this type of soil is most likely:

Control

i. Plant resistant varieties if nematodes are known to be present in the soil.

ii. Check roots of plants mid-season or sooner if symptoms indicate nematodes.

iii. Soil solarisation and fumigation.

2. Armyworm (*Pseudaletia unipuncta*)

Singe or closely grouped circular to irregularly shaped holes in foliage. Sometimes due to heavy feeding of foliage by young larvae leads to skeletonized leaves. Dry wounds are also observed on fruit. The worm lays eggs on the leaves. Egg clusters of 50-150 eggs may be present on the leaves and are covered in a whitish scale which gives the cluster a cottony or fuzzy appearance.

Control

i. Biological control by natural enemies which parasitize the larvae and the application of *Bacillus thuringiensis*.

3. Cutworms *Agrotis* spp., *Peridroma saucia, Nephelodes minians* and others

The larvae cause the damage and are usually active at night and hide during the day in the soil at the base of the plants or in plant debris of toppled plant. Cutworms have a wide host range and attack vegetables including asparagus, bean, cabbage and other crucifers, carrot, celery, corn, lettuce, pea, pepper, potato and tomato.

Control

i. Remove all plant residue from soil after harvest or at least two weeks before planting, especially if the preceding crops are alfalfa, beans or a leguminous cover crop.

ii. Cover the bottom 3 inches of the stem above the soil line and a couple of inches into the soil with plastic or foil collars fitted around plant stems can prevent larvae severing plants.

iii. Application of insecticides to infested areas.

4. Aphids (Willow-carrot aphid) *Cavariella aegopodii*

Small soft bodied insects on underside of leaves and/or stems of plant; usually green or yellow in color; if aphid infestation is heavy it may cause leaves to yellow and/or distorted, necrotic spots on leaves and/or stunted shoots; aphids secrete a sticky, sugary substance called honeydew which encourages the growth of sooty mould on the plants.

Control

i. Pruning of affected leaves and stem.

ii. Use tolerant varieties if available.

iii. Use of reflective mulches such as silver coloured plastic can minimize aphids from feeding on plants.

iv. Spray insecticides when infestation is very high. Plants generally tolerate low and medium level infestation; insecticidal soaps or oils such as neem or canola oil are usually the best method of control.

Diseases

1. Powdery mildew (*Erisyphe heraclei*)

Powdery growth on leaves, petioles flowers stalks and bracts; leaves becoming chlorotic; severe infections can cause flowers to become distorted. Disease emergence is favoured by high humidity and moderate temperatures; infection is most severe in shaded areas.

Control

i. Plant tolerant varieties.

ii. Avoid excess fertilization.

iii. Spray of sulphur fungicides. Sulfur application can be used in infection occurs early in season.

2. Damping-off *Pythium* spp., *Rhizoctonia solani*

Soft, rotting seeds which fail to germinate; rapid death of seedling prior to emergence from soil; collpase of seedlings after they have emerged from the soil caused by water-soaked reddish lesions girdling the stem at the soil line.

Control

i. Avoid planting dill in poorly drained, cool and wet soil.

ii. Planting in raised beds will help with soil drainage.

iii. Treat seeds with fungicide prior to planting to eliminate fungal pathogens.

3. Cercospora leaf blight *(Cercosporidium punctum)*

Small, necrotic flecks on leaves which develop a chlorotic halo and expand into tan brown necrotic spots; lesions coalesce and cause leaves to wither, curl and die.

Control

i. Plant only pathogen-free seed.

ii. Crop rotation should be followed.

iii. Apply appropriate fungicide sprays.

Varieties

1. Ajmer dill-1 (NRCSS-AD-1) –It has been released from NRCSS, Ajmer. It is a European type dill variety which is suitable for cultivation under irrigated condition. It has good content of herbal constituents. The leaves are dark green in colour with average. Plant height of 134 cm and yield 14.7q/ha of seed.
2. Ajmer dill-2 (NRCSS-AD-2)–Also released from NRCSS, Ajmer. It is Indian type dill variety with bold, compact and dark brown seeds. It is suitable for irrigated and rainfed condition. Average plant height is 90cm, seed yield is 14.6q/ha under irrigated condition.5.8 q/ha under rainfed condition, seed contain 3.2% essential oil.

5.3 OPIUM POPPY

Botanical Name	*Papaver somniferum* L.
Family	Papaveraceae
Origin	Western Mediterranean region
Chromosome number	2n=22
Commercial value	Leaves. Seeds, alkaloids from latex
Chemical constituents	Papaverine, thebain

Research Institute	i. Central Institute of Medicinal and Aromatic Plants, Lucknow.
	ii. National Botanical Research Institute, Lucknow.
	iii. JNKVV, Jabalpur.
	iv. RVSKVV, Gwalior.
	v. Narendra Dev University of Agriculture and Technology, Faizabad.
	vi. National Bureau for Plant Genetic Resources, New Delhi.
	vii. Rajasthan Agricultural University, Udaipur.

Importance

Opium poppy has its importance as a plant based natural pain reliever from the time dating back to early civilization till today. Its pain relieving properties had been described in various books of unani, allopathy and ayurvedic medication system. It contains five major alkaloids *viz.,* morphine, codeine, thebaine, noscapine and papaverine which are present in opium latex in ample amount. The opium poppy is the source of two food ingredients: poppy seed and poppy seed oil. The seeds contain very low levels of opiates, and the oil extracted from them contains even less. Both the oil and the seed residue also have commercial uses. The seeds of opium poppy are highly nutritious as it contains protein upto 24% and other vital nutrients beneficial for human health. The leaves of the plant are used as vegetable in some places in the world. The seed oil of poppy is also important for health point of view due to having high percentage of linoleic acid (68%) which helps in lowering blood cholesterol level in human body and is also used in the treatment of cardiovascular diseases in human system [Vos and Cunnne,2003; Sacks and Campos, 2006].

Australia, Turkey and India are the major producers of poppy for medicinal purposes and poppy-based drugs, such as morphine or codeine. The USA has a policy of sourcing 80% of its narcotic raw materials from the traditional producers, India and Turkey.

Distribution

The opium poppy is distributed in the temperate and subtropical regions of the old world extending from 60°North West Soviet Union whereas the southern limit reach almost the tropics. Legally it is cultivated in India, China, Russia, Egypt, Yugoslavia, Czechoslovakia, Poland, Germany, Netherland, Japan,

Argentina, Spain, Bulgaria, Hungary and Poland [Singh *et al.*, 1995]. India is the largest opium producing and exporting country in the world. In India its cultivation is confined to the states of Madhya Pradesh, Rajasthan and Uttar Pradesh.

Botany

Opium poppy is the only species of Papaveraceae that is an agricultural crop grown on a large scale. Other species, *Papaver rhoeas* and *Papaver argemone*, are important agricultural weeds. *Papaver somniferum* Paeoniflorum group (sometimes called *Papaver paeoniflorum*) is a subtype of opium poppy whose flowers are highly double, and are grown in many colors. *P. somniferum* Laciniatum group (sometimes called *Papaver laciniatum*) is a subtype of opium poppy whose flowers are highly double and deeply lobed, to the point of looking like a ruffly pompon.

Description of Plant

Poppy is a small, erect, scarcely branched (towards the top) herb attaining a height of 120cm with large, serrated leaves and attractive bright coloured flowers.the main shoot and branches terminate into large, oblong to globose capsules, filled with small white, flat seeds. The seeds are small, kidney shaped, grey- brown, reddish brown or black and rich in oil.

Soil

It prefers a well drained, highly fertile, light or loam soil with an optimum pH of around 7.0.

Climate

It is basically a temperate climate crop but can be grown successfully during winter in, sub tropical regions. The poppy crop requires a maximum temperature upto 20°C at the time of germination while dry weather at the maturity. Cool climate favours higher yield, while, higher day/night temperature generally affects the yield. Frosty condition, cloudy or rainy weather tends to reduce not only the quantity but also the quality of opium.

Cultivation

Poppy seeds should be sown in very well prepared soil. The field should be given 5-6 cross- ploughings, followed by planking. The land should be divided into small plots to facilitate irrigation.

Propagation

It is propagated through seeds. Opium poppy is one of the cultivated plants with highest propagation rate because fruits contain numerous seeds. Seeds are very minute with 1000 seeds weighing only 0.5g.

Sowing time

The best time for sowing of opium poppy is late October or early November.

Seed rate and Spacing

Seed rate depends upon method of sowing for broadcast method. It ranges from 7-8 kg/ha and for line sowing it is 4-5 kg/ha. Spacing of 30 cm between lines and 30 cm between plants is normally adopted.

Manuring and Fertilization

Farm yard manure @ 20-30 t/ha is generally applied at the time of field preparation. Besides, 60-80 kg of N and 40-50 kg of P_20_5 per ha is recommended. No potash is applied. Half of N and entire P are applied at sowing time through placement and remaining half of N placed at rosette stage.

Intercultural operations

Irrigation

Water management is very important part of poppy cultivation. Light irrigation is given immediately after sowing. Then again the field is irrigated after 7 days when the seeds start germinating. Till pre-flowering stage three irrigations at an interval of 12-15 days as the weather conditions prevails. The irrigation frequency should be reduced at 8-10 days during flowering and capsule formation stage.

Weeding

Weeding and hoeing are also necessary for providing poppy seedling a better chance to grow. The first weeding is done 20-25 days after sowing followed by 15-20 days interval. For good yield optimum spacing between the plants should be maintained at a distance of 10cm apart.

Lancing and Latex Collection

Opium starts flowering in 95-115 days after sowing. The petals start shedding after 3-4 days of flowering. The capsules mature after 15-20 days of flowering.

Lancing of the capsules exudes maximum latex at this stage. This stage can be visually judged by the compactness and a change in the colour from greenish to light green coloured ring in the capsule. The stage is called as industrial maturity. Lancing is done by cutting of the superficial layer of the capsule wall from which the latex oozes out. The mature green capsule is lanced with an instrument called "Nastar"having 3-4 small blades designed to ensure uniformity in depth of incisions. Generally 3-4 lancing is done in each capsule with parallel longitudinal cuts which is performed after mid day and allow the latex to remain overnight on the capsules for coagulation. In the following morning the latex is collected from the capsule walls with blunt edge of small iron scoop. The opium is kept in small plastic box or earthen pot or copper bowls. The latex colour varies from dark to light brown to pink based on the variety. The depth of incision should not be more than 1.2 mm. After collection of opium, lanced capsules are left to dry over plants for next 15-20 days for harvesting of seeds (Mishra *et al.*, 2013).

The alkaloids are mainly extracted in India from green unripe capsules by making incision upto 1-2 mm in the epidermal wall of the capsule, but globally it is extracted from the dried capsule which is called CPS (Concentrated poppy straw) system. In CPS system, the dried capsules along with eight inches of peduncle are harvested and seeds are threshed. The remaining husk is used to extract various alkaloids. The whole plant parts of opium poppy are valuable in terms of food, medicine, vegetable and as brewages.

Harvesting

The capsules, after the lancing operation and collection of opium latex, are allowed to dry on the plant itself. The drying process takes about 15 days after the lancing is completed.In India, the capsules are plucked by hand and the seeds are separated after breaking the capsules. A dry capsules weigh about 7 g and it contains 11-12 thousand seeds weighing about 3.5 to 4 g.

Yield of crude opium and seed

On an average 25-30kg/ ha of crude opium and 400-500kg/ha of seeds are obtained in India.

Plant protection

Insect-pests

1. Root weevil (causal agent-*Sternocarus fuliginosus*)

This pest causes maximum damage to poppy crop. It bores into upper parts of the roots which ultimately turns blackish and leaves wither due to chlorosis. The larva mines the leaf lamina and damages leaves.

2. Cutworm (causal agent - *Agrotis* spp.)

The larva of this pest is active at night and remains hidden in cracks in the ground. It mostly affects the young plants by destroying basal part of the stem.

Control

i. It include hand picking of the caterpillars and dusting the crop with 2 % carbaryl.

3. Aphid (causal agent - *Myzus persicae)*

This is also another major pest of opium poppy crop. The nymphs and adults suck the leaf sap results damage of leaf and consequently whole plant.

Control

i. The control measures include spray of neem oil 0.5%.

ii. Natural enemies like coccinellid beetle can also be encouraged.

4. Capsule borer (causal agent - *Helicoverpa armigera)*

The capsule borer is also a serious pest in opium poppy which harms capsule to a maximum extent. they destroys whole capsule eating up the floral head and seeds.

Control

i. Hand picking of the larvae along with pheromone traps is recommended while spray of NPV 250 l//ha is also beneficial.

ii. The use of Bt spray formulation @ 2g/2ml per litre of water and use of egg parasitoids *Trichogramma chilonis* @ 5cc/ha is also found effective.

Diseases

1. Downy Mildew (CO- *Peronospora arborescens*)

The infection starts spreading upwards from the lower leaves and the entire leaf surface gets covered by brown powder. The plants dies prematurely as the stem, branches and even capsules are also attacked by fungus. Capsule formation is also adversely affected due to infection causing significant reduction in opium yield.

Control

i. Disinfection and spraying of the seed beds with 0.5% Bordeaux mixture and different copper containing fungicides.

2. Powdery Mildew (CO- *Erysiphae polygoni*)

This disease is causes severe losses in opium production. The symptoms appear in late stages of plant growth with white powder on the surface of leaves and capsules.

Control

i. It includes field sanitation along with spray of *Spersul* (0.5%) and seed disinfection.

3. Collar Rot disease (CO- *Rhizoctonia solani* Kühn.)

This is one of the most severe fungal diseases of opium poppy caused by, premature death of infected plant and ultimately Decline in seed yield.

4. Leaf Blight (causal agent -*Pleospora calvescens*)

Defused yellow spots on leaves followed by premature drying of infected leaves. High temperature and heavy rainfall favours the disease.

Varieties

Sujata	This variety has been developed by Central Institute of Medicinal and Aromatic Plants, Lucknow. In this variety plants are 80-100 cm tall having 3-4 flat glabrous capsules with 18-20 cm long erratic black peduncle. It is an opium less, alkaloid less and non narcotic variety. This variety assist opium cultivators to grow food grade poppy without any restriction or permission in form of license.

Subhra	This variety has been developed by Central Institute of Medicinal and Aromatic Plants, Lucknow. In this variety plants are medium sized having 3-4 capsules/plant. The average seed yield and husk yield of the variety are approx. 910kg/ha and 790kg/ha respectively. The morphine percentage in husk is approx. 0.77%.
Sweta	This variety has been developed by Central Institute of Medicinal and Aromatic Plants, Lucknow. It is high yielding variety with 66.5kg/ha opium yield and about 18% morphine content in latex.
BROP-1	In this variety the plants are medium sized having 3-4 capsules/plant and capsules are of three types viz. oily, parrot coloured and black peduncle. Flowers are white. It is a synthetic variety stabilized after hybridization/intermating between three high yielding cultivars viz. kali dandi (black peduncle), suga pankhi (parrot color) and sufaid dandi (white peduncle) followed by selection. The average opium yield, seed yield and morphine content are up to 54kg/ha, 1000-1200kg/ha and 13% respectively.
NBRI-1	This variety is developed through selection. The plants are medium tall havinglarge fringed leaves and white flowers. The average opium yield, seed yield and morphine content are up to 52kg/ha, 1000kg/ha and 12-13% respectively. It is suitable for cultivated mainly in Northern Central India.
Jawahar Aphim 16 (JA-16)	Developed by Jawaharlal Nehru Krishi Vishwavidalaya, College of Agriculture, Jabalpur (M.P.). It is a downy mildew resistant variety by pure line selection. Plants of this variety are tall having white flowers and 1-3 big capsules/plant. The variety has opium yield, seed yield and morphine content up to 45-54kg/ha, 900-1000kg/ha and 12% respectively. It is recommended for cultivation in Madhya Pradesh.
Kirtiman (NOP-4)	This variety has been developed at Narendra Dev University of Agriculture and Technology, Faizabad. It is downy mildew resistant variety by selection. The plants of this variety are quite tall having white flowers and 1-2 oval capsules. Opium yield, seed yield and morphine content of the variety ranges 35-46kg/ha, 900-1100kg/ha and up to

	12% respectively. Eastern U.P. region is best suited for its cultivation.
Chetak (UO 285).	Developed by Rajasthan Agricultural University, Udaipur and is resistant to disease and lodging. The plants of this variety are average tall with big capsules. Flowers are white with smooth petals. Opium yield, seed yield and morphine content ranges up to 54kg/ha, 1000-1200kg/ha and 12% respectively. It is most suitable for cultivation in Rajasthan.

(*Source*: Mishra *et al.*, 2013)

References

Mishra, BK, Rastogi, A., Ameena Siddiqui, Srivastava, M., Verma, N., Pandey,R. Sharma NC and Shukla, S. (2013). Opium Poppy: Genetic Upgradation Through Intervention of Plant Breeding Techniques In: Chapter 8 Plant Breeding from Laboratories to Fields. In Tech : 209-239.

Vos E and Cunnne SC. (2003). '-Linolenic acid, Linoleic acid, Coronary heart diseases and overall mortality. *American Journal of Clinical Nutrition*. 72(2): 521-522.

Sacks FM and Campos H. (2006) Polysaturated fatty acids, inflammation and cardiovascular diseases: Time to widen our view of the mechanism. *Clinical Endocrinology Metabolism*. 91(2): 398-400.

Singh SP, Khanna KR, Shukla S, Dixit BS and Banerjee R.(1995) Prospects of breeding opium poppies (*P. somniferum* L.) as a high linoleic acid crop. *Journal of Plant Breeding*. 114: 89-91.

5.4 ASHWAGANDHA

Botanical Name	*Withania somnifera* (L.) Dunal
Family	Solanaceae
Chromosome No	2n = 48
Distribution	India, North Africa
Chemical constituents	Withanine, somniferine

Importance

It is a well known as "wonder herb" due to Ayurvedic, Unani and indigenous medicinal properties. It is widely used in several indigenous systems of medicine for the treatment of various ailments, *viz*. asthma, bronchitis, inflammatory diseases, and ulcer and stomach problems. Ashwagandha is used for white leprosy, cough, asthma, uterive. Diseased root is used in application in obstinate ulcers and rheumatic swelling. Withania Somnifera is used as a nutrient and health restorative to pregnant and old people. A decoction of the root is recommended or its powder with milk may be taken.

Botany

It is commonly known as Ashwagandha which indicates the equine (of horses) odour of the plant. This plant is an erect herbaceous, evergreen shrub of 13 to 150 cm (4 to 5 ft). The stems are covered with whitish colour, stellate hairs. The leaves are ovate entire, thin. The flowers ate bisexual garnish or cured yellow, axillary in clusters of about 25 forming cymes, sessile.

Soil

It can be grown under various type of soil but it grows well under sandy loam or laterite soil with good condition of soil and having proper drainage.Loose soil facilitate the growth of roots as well as digging is also easy without damaging the roots. Most suitable soil pH for proper growth is 7.5-8.

Climate

It grows well under the subtropical climatic condition and prefers dry weather it is planted during the late rainy season.

Method of Sowing

It can be grown by two methods:

A) Direct Sowing

After receiving the 2-3 rain showers, the field is prepared and divide into plots. seeds can be sown directly into the field.

Time of sowing- seeds are directly sown during the second week of July.

Seed rate- 5 kg/acre.

B) Nursery Raising

About 2kg of seeds are required for 1 acre. To avoid diseases; the seeds are treated with Dithane M-45 at the rate of 3g per kg of seed before sowing. The seeds in nursery beds are sown in lines spaced at 5cm and covered with light soil or sand. Its starts to germinate within 7 to 8 days. When the seedling grows to a height of 60cm that is in 6 weeks time, they are transplanted in row of the main field. This crop does not require heavy doses of manure and fertilizers. Hand weeding should be done at intervals of 30 days to control the weeds.

Interculture Operation

The directly sown crop is thinned at 25-30 days to maintain a plant population of 20,000-25000/ha. Hand weeding at 30 days interval helps to control the weeds effectively.

Manures and Fertilisers and Pesticides

The medicinal plants have to be grown without chemical fertilizers and use of pesticides. Organic manures like, Farm Yard Manure (FYM), Vermi-Compost, Green Manure etc. may be used as per requirement of the species. To prevent diseases, bio-pesticides could be prepared (either single or mixture) from Neem (kernel, seeds & leaves), Chitrakmool, Dhatura, Cow's urine etc. should be used.

Irrigation

Light shower after transplantation ensures establishment of seedlings. There is no need of irrigation if rainfall is at regular intervals. Excessive rainfall/water is harmful to the crop. Life saving irrigations may be applied, if required.

Plant Protection

Diseases

Seedling blight, leaf blight are the common diseases effecting Ashwagandha. Incidence can be minimized by treating the seeds before sowing with captan @3g/kg of seeds followed by spraying of the crop with Dithane M-45 @ 3g /litre, when the crop is 30 days old. The spraying should be repeated at the interval of 7 to 10 days if the disease is not controlled.

Harvesting

After 8 months of planting the maturity of the crop is judged by the drying out of leaves and berries turning red. The whole plants are uprooted and the roots are separated from the aerial part by cutting the stem 1-2 cm above the crown. They are then transversely cut into small pieces of 7 to 10 cm for drying. The berries are plucked from the dried plant and are threshed to obtain the seeds.

Yield

An average yield of 300-500kg/ha of dried roots and 50-75kg/ha of seeds can be obtained.

Grading

On the basis of the thickness and uniformity of the tap root pieces it can be graded into three grades:

A- Grade: Root pieces upto 7 cm in length, solid with 1-1.5cm diameter, they should be brittle and pure white on the inside.

B- Grade: Root pieces upto 5cm in length, solid with a diameter of less than 1cm, the roots should be brittle and white on the inside.

C- Grade: Root pieces upto 3-4cm in length, side branches solid, with a diameter of 1 cm or less.

5.5 GUGGAL OR INDIAN BEDELLIUM TREE

Botanical name	*Commiphora mukul*
Chromosome No.	2n=22
Family	Burseraceae
Origin	Africa and Asia

Importance

It is a source of Indian Bedellium, a oleo-gum resin (mixture of the resin) obtained by incision of the bark. It is highly in the treatment of obesity, diabettis mellitus,arthritis,indolent ulcer, weak and spongy gums, chronic dyspepsia. *Guggal* removes three *doshas* like *kapha*, *pitta* and *vata*. Guggul stimulates *pitta* and thus enhances warmth, digestion, circulatory and reproductive processes. It also regulates *vata* (nerve force) and *kapha* (fluidic aspects).

The Sanskrit definition of the term "guggul" is "one that protects against diseases." This attests to the wide respect and therapeutic Ayurvedic applications for this botanical, considered the most important for the removal of "ama," toxic substances which accumulate as a result of sluggish digestion and circulation associated with a slowing of metabolism. Guggul is warming and stimulates metabolism that is why it is one of the few botanicals that has been shown to treat hypothyroid conditions.

Distribution

It is found in the arid to semi-arid areas of Northen India, Bangladesh and Pakistan.

Botany

It is a tree or shrub upto 3-4m high, the brances are aromatic and end in sharp spines. Plants are dimorphic, one having bisexual and male flowers and the other having female flowers with staminodes.

Soil

It is not grown on a commercial plantation scale anywhere.But the plants are found growing naturally in arid tracts of western India over sandy to silt loam soils, which are poor in organic matter but rich in minerals. Faster growth of the plant is observed in soil which have moisture retaining capacity. An average soil is suitable for its cultivation.

Climate

The plant has wide adaptability and is found growing in arid regions. Its small, leathery leaves and thick bark covered with a white waxy coat over the stem, helps it to withstand desert conditions. It prefers a warm, dry climate for a good yield of oleo-gum-resin.

Land Preparation

The land is prepared well in advance of the rainy season by 2-3 ploughing and laid out into plots of convenient sizes. Pits of the size 0.5x 0.5x0.5m. are dug at a spacing of 3-4m, in rows.The pits are filled with FYM and topsoil and mixed with Aldrin (5%) to prevent and damage to the plants due to termite attack.

Cultivation

Propagation

Guggal can be propagated by seeds and vegetatively through stem-cuttings.Air layering also successful.

i) Propagation by seeds- this method is not a common method of propagating the plant.because the seed germination is slow and very poor (5%) due to the presence of the hard seed coat. For achiving a good rate of germination, seeds are mechanically scarified with sand paper and are kept under running water for 24 hours. The seedlings can be raised in polythene bags during the *kharif* season and, after hardening, may be planted in the main field.

ii) Propagation by stem cutting: About 15-20cm long and 10mm thick, semi-hardwood cuttings one taken and cut end is treated with IBA or NAA growth regulator solution and planted in well prepared and manure nursery beds during the month of June and July. Irrigation is given after planting and regularly thereafter. The cutting sprouts in 10-15 days, they grow well and are ready for planting in the main field after 10-12 months, during the next rainy season. The percentage of rooting in stem cutting is around 80-94%.

Planting

The rooted cuttings are planted in the pits during the rainy season. As the plants grow, they are trained properly by cutting the side branches.

Fertilizer Application

This crop not showed positive response to fertilizers, except to low level of nitrogen fertilization. Hence, urea or ammonium sulphate at 25/50g/bush is given twice a year before irrigation.

Irrigation

Providing light irrigation during the summer season supports good growth of the plants.

Interculture operation

Intercultural is confined to one weeding and hoeing in the early stages of growth. But the soil has to be stirred around the bushes twice in a year. This practice of soil stirring is found to increase the growth of plants.

Pests and Diseases

Guggal plants are attacked by a leaf eating caterpillar *Euproctis lantana* Walker; White fly- *Bemisia tabaci*, which sucks the sap of leaves as a result of which the leaves turn yellow and finally drop off. The rooted cuttings also suffer from termite attacks in the early stage of the growth. The diseases notices on this crop are leaf spot caused by *cercospora* spp, and bacterial leaf blight.

Gum-tapping and yield

The plants attain normal height and girth after 8-10years of growth, when they are ready for tapping. For tapping the gum, which is present in the balsam canals in the phloem, bark deep, i.e. a shallow incision is made on the bark. If the incision is too deep, the plant either dies or yields little resin in the following years. While making the incision, a small quantity of guggual gum mixed with water may be applied to the incised place using a prick-chisel.The yield of gum resin from each plant is about 700-900g.

Varieties

Marusudha

Highest yielder and released from Gujarat for cultivation.

Appendices

Appendix 1: Important flavour compounds from major spices

Spice	Important flavoring compound
Allspice	Eugenol, â-caryophyllene
Anise	(E)-anethole, methyl chavicol
Black pepper	Piperine, S-3-Carene, â-caryophyllene
Cardamom	a-terpinyl acetate, 1-8-cineole, linalool
Turmeric	Curcumin,Turmerone, Zingiberene, 1,8-cineole
Ginger	Gingerol, Shogaol, Zingiberene, geranial
Mace	sabinene, Myristicin,1-terpenin-4-ol
Nutmeg	Sabinine, myristicin,elemicin
Cumin	Cuminaldehyde, p-l,3-mentha-dienal
Fennel	(E)-anethole, fenchone
Saffron	Safranol
Vanilla	Vanillin, p-OH-benzyl-methyl ether

Appendix II: Plant part used for major spices

Parts	Crops
Floral parts	Clove, Saffron
Fruits	Cardamom, Chilli All spice, Black pepper
Seed	Aniseed, Caraway
Rhizomes	Ginger, Turmeric
Roots	Angelica, Horse-radish
Leaves	Bay leaves, Mints
Kernel	Nutmeg
Bark	Cinnamon, Cassia
Bulb	Garlic, Onion

Appendix III: Products and byproducts of spices

Spices	Products
Cardamom (large)	Oil, oleoresin
Cardamom (small)	Oil, oleoresin, green cardamom
Black pepper	Oil, oleoresin, green pepper in brine, dehydrated green pepper, freeze dried green pepper, frozen green pepper, white pepper, ground pepper, organic pepper, sterile pepper, canned tender green pepper
Turmeric	Dehydrated turmeric powder, curcuminoids, oil, oleoresin
Ginger	Oil, oleoresin, candy, preserves, various forms of ginger powder, ginger brandy, wine, beer, medicinal beverages, encapsulated ginger oil, dehydrated ginger
Chillies	Chilly powder, green chilly in brine, canned green chilly, chilly paste, pickles, capsicum iol and oleoresins.
Coriander	Coriander powder, organic coriander, dried coriander leaves, Oil, oleoresin.
Cumin	Oil, oleoresin, cumin powder, medicinal beverages, cumin sips, organic cumin.
Fennel	Oil, oleoresin,fennel powder, sugar coated funnel whole seed, organic fennel, fennel brandy, wine and beer, medicinal beverages, fennel sip.
Fenugreek	Fenugreek powder, organic fenugreek, dried fenugreek leaves

Appendix IV: Products and byproducts of plantation crops

Crops	Products and byproducts
Coconut	Food, fuel, cosmetics, timber, husk, copra, coconut water, coconut oil, coconut kerenel, coconut sweets,milk poder.
Cashewnut	Sweets, cosmetics,cashew liquid, cashew shell, cake,cashew apple fenni
Cocoa	Coca nibs,cocoa mass,cocoa powder,cocoa butter,chocolate
Palm oil	Palm oil
Palmyrah palm	Neera